STRATHCLYDE UNIVERSITY LIBRARY
KV-038-756

TREATMENT OF MICROBIAL CONTAMINANTS IN POTABLE WATER SUPPLIES

TREATMENT OF MICROBIAL CONTAMINANTS IN POTABLE WATER SUPPLIES

Technologies and Costs

by

Jerrold J. Troyan
Sigurd P. Hansen

CWC-HDR, Inc.
Cameron Park, California

NOYES DATA CORPORATION
Park Ridge, New Jersey, U.S.A.

Library of Congress Catalog Card Number: 89-16034
ISBN: 0-8155-1214-7
ISSN: 0090-516X
Printed in the United States

Published in the United States of America by
Noyes Data Corporation
Mill Road, Park Ridge, New Jersey 07656

10 9 8 7 6 5 4 3 2 1

Library of Congress Cataloging-in-Publication Data

Troyan, Jerrold J.
Treatment of microbial contaminants in potable water supplies : technologies and costs / by Jerrold J. Troyan, Sigurd P. Hansen.
p. cm. -- (Pollution technology review, ISSN 0090-516X ; no. 171)
Bibliography: p.
Includes index.
ISBN 0-8155-1214-7 :
1. Drinking water--Purification. 2. Microbial contamination. I. Hansen, Sigurd P. II. Title. III. Series.
TD433.T76 1989
628.1'62--dc20 89-16034
CIP

Foreword

This book identifies the best technologies or other means that are generally available, taking costs into consideration, for inactivating or removing microbial contaminants from surface water and groundwater supplies of drinking water. For municipal officials, engineers, and others, the book provides a review of alternative technologies and their relative efficiency and cost. More specifically, it discusses water treatment technologies which may be used by community and noncommunity water systems in removing turbidity, *Giardia,* viruses, and bacteria from water supplies. While most of the book covers surface water supplies, a brief discussion of disinfection technologies and costs for groundwater supplies is also provided, since disinfection is the best available technology for groundwater systems to comply with coliform regulations.

The technologies and actions available to a community searching for the most economical and effective means to comply with microbiological regulations include modification of existing treatment systems; installation of new treatment systems; selection of alternate raw water sources; regionalization; and documenting the existence of a high quality source water while implementing an effective and reliable disinfection system, combined with a thorough monitoring program, and maintaining a continuing compliance with all drinking water regulations.

It is not the intent of the USEPA to require any system to use a particular technology to achieve compliance with proposed treatment regulations. Instead, the responsibility is retained by the individual water systems to select one or more procedures that are optimal for their particular water supply situation. Whatever technology is ultimately selected by a water supplier to achieve compliance with the requirements must be based upon a case-by-case technical evaluation of the system's entire treatment process, and an assessment of the economics involved. However, the major factors that must be considered include:

- Quality and type of raw source water
- Raw water turbidity
- Type and degree of microbial contamination
- Economies of scale and the potential economic impact on the community being served
- Treatment and waste disposal requirements

Some methods are more complex or more expensive than others. Selection of a technology by a community may require engineering studies and/or pilot-plant operations to determine the level of removal a method will provide for that system.

Alternative technologies for the removal of microbial contaminants and turbidity are identified because of their adaptability to treatment of drinking water supplies. It is expected that after development and pilot-scale testing, these methods may be technically and economically feasible for specific situations.

The information in the book is from *Technologies and Costs for the Treatment of Microbial Contaminants in Potable Water Supplies,* by Jerrold J. Troyan and Sigurd P. Hansen of CWC-HDR, Inc. for the U.S. Environmental Protection Agency, October 1988.

The table of contents is organized in such a way as to serve as a subject index and provides easy access to the information contained in the book.

Advanced composition and production methods developed by Noyes Data Corporation are employed to bring this durably bound book to you in a minimum of time. Special techniques are used to close the gap between "manuscript" and "completed book." In order to keep the price of the book to a reasonable level, it has been partially reproduced by photo-offset directly from the original report and the cost saving passed on to the reader. Due to this method of publishing, certain portions of the book may be less legible than desired.

Acknowledgments

Preparation of this document involved important contributions from many people in two consulting engineering firms and the United States Environmental Protection Agency (USEPA), Office of Drinking Water, and Drinking Water Research Division. In fulfillment of a contract with the USEPA, day-to-day work was conducted by CWC-HDR, Inc., with supervision, review, and technical contributions provided by Malcolm Pirnie, Inc.

Personnel from Malcolm Pirnie involved in this work were John E. Dyksen, David J. Hiltebrand, and Linda L. Averell. Principal authors from CWC-HDR were Jerrold J. Troyan and Sigurd P. Hansen. Other members of CWC-HDR (which became HDR Engineering, Inc., in April 1989) who contributed to either the technical content or the preparation of the manuscript include:

Teresa D. Boon
May L. Bray
Candice E. Cornell
Gordon L. Culp
Russell L. Culp
Brian A. Davis
Marie A. Filippello
Perri P. Garfinkel
Judith A. Hinrichs
Robert R. Livingston
Mark S. Montgomery, Ph.D.
I. Jean Wagy
Bruce R. Willey

In addition, valuable technical review and contributions to the text were provided by USEPA personnel including Stig Regli, Project Manager, Gary S. Logsdon, John C. Hoff, Ph.D., Edwin E. Geldreich, and James J. Westrick.

NOTICE

The materials in this book were prepared as accounts of work sponsored by the U.S. Environmental Protection Agency. On this basis the Publisher assumes no responsibility nor liability for errors or any consequences arising from the use of the information contained herein. Mention of trade names or commercial products does not constitute endorsement or recommendation for use by the Agency or the Publisher.

Final determination of the suitability of any information or procedure for use contemplated by any user, and the manner of that use, is the sole responsibility of the user. The reader is warned that caution must be exercised when dealing with potentially hazardous materials such as contaminated waters, and expert advice should be sought at all times before implementation of any treatment technologies.

Contents and Subject Index

Executive Summary

INTRODUCTION (SECTION I)

This document assists the Administrator of the U.S. Environmental Protection Agency (EPA) in identifying the best technologies or other means that are generally available, taking costs into consideration, for inactivating or removing microbial contaminants from surface water and groundwater supplies of drinking water. For municipal officials, engineers, and others, the document provides a review of alternative technologies and their relative efficiency and cost. More specifically, this document discusses water treatment technologies which may be used by community and noncommunity water systems in removing turbidity, Giardia, viruses, and bacteria from water supplies. EPA is currently developing treatment regulations addressing these microbiological concerns. While most of this document is devoted to discussion of surface water suppiles, a brief discussion of disinfection technologies and costs for groundwater supplies is also provided, since disinfection is the best available technology for groundwater systems to comply with the coliform regulations.

It is not the intent of EPA to require any system to use a particular technology to achieve compliance with the proposed treatment regulations. Instead, the responsibility is retained by the individual water systems to select one or more procedures that are optimal for their particular water supply situation. Whichever individual or combination of technologies is ultimately selected by a water supplier to achieve compliance with the requirements must be based upon a case-by-case technical evaluation of the system's entire treatment process, and an assessment of the economics involved.

The information provided in the main document is intended to aid a system in reviewing available technologies for achieving the required reduction in turbidity and microorganisms. It provides the user with an evaluation of the various methods in use today for the removal of different concentrations of turbidity and microorganisms, as well as relative costs.

Definition of Technology Categories

The methods that can be applied for the removal of microbial contaminants are divided into three categories:

Most Applicable Technologies: Those that are generally available and have a demonstrated removal or control based on experience and studies for most systems subject to the regulations, and for which reasonable cost estimates can be developed.

Other Applicable Technologies: Those additional methods not identified as generally available, but which may have applicability for some water supply systems in consideration of site-specific conditions, despite their greater complexity and cost.

Additional Technologies: Those experimental or other methods with potential use that may be studied for specific situations to achieve compliance with the regulations, and for which insufficient data exist to fully evaluate the suitability and applicability of the technology for removal of microbial contaminants.

BACKGROUND (SECTION II)

An overview of trends in the incidence of waterborne disease over the past 40 years, and of the capabilities and limitations of available treatment processes, indicates the need for: (1) constant awareness of a broad spectrum of disease-producing organisms; (2) continuing improvement in microbial detection techniques; and most importantly, (3) proper application and operation of available treatment processes.

Waterborne Disease Outbreaks

Annual occurrences of outbreaks of waterborne diseases from 1946 to 1980 are shown on Figure 1. The increase in annual outbreak occurrences since 1966 is

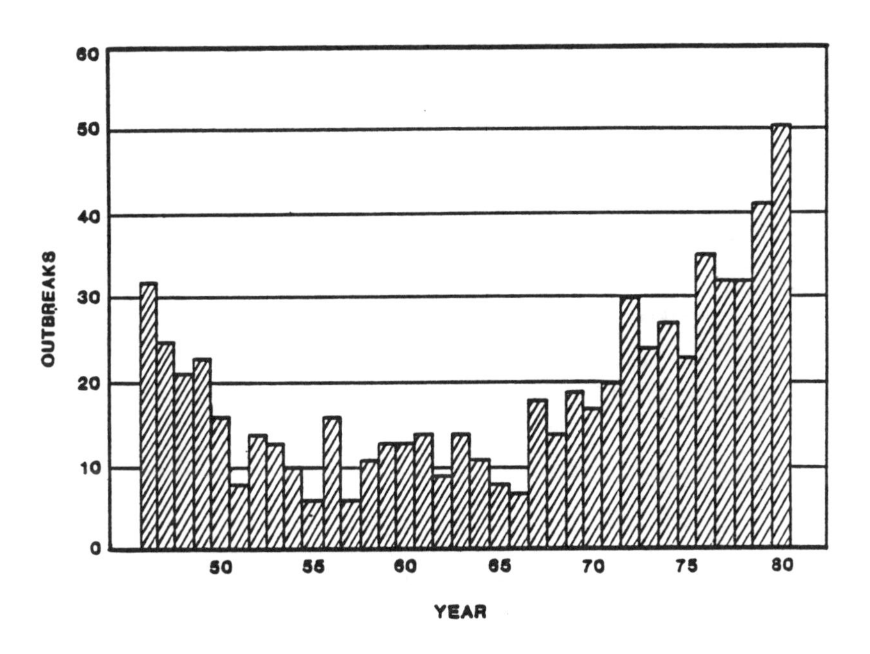

Figure 1
ANNUAL OCCURRENCE OF WATERBORNE DISEASE OUTBREAKS

probably due to both more active data collection by federal agencies, and more aggressive investigating and reporting by a few states.[1] The most recent data still lack accuracy due to the absence of intensive surveillance by many state and local agencies.

Reported causative agents for waterborne diseases from 1972 to 1981 are indicated in Table 1. Giardia ranks number one in cases of illness and outbreaks.

TABLE 1. ETIOLOGY OF WATERBORNE OUTBREAKS IN THE UNITED STATES, 1972-1981

	Outbreaks	Cases
Acute gastrointestinal illness	183	37,069
Giardiasis	50	19,863
Chemical poisoning	41	3,717
Shigellosis	22	5,105
Hepatitis A	10	282
Salmonellosis	8	1,150
Viral gastroenteritis	11	4,908
Typhoid fever	4	222
Campylobacter gastroenteritis	4	3,902
Toxigenic E. coli gastroenteritis	1	1,000
Cholera	1	17
TOTAL	335	77,235

Source: Reference 2.

An evaluation of water system deficiencies specifically responsible for giardiasis outbreaks is presented in Table 2. It is worth noting that 83.5 percent of the cases listed in Table 2 (first three items, surface water systems) occurred in systems that either did not have filtration, or in which filtration was ineffective or intermittent. Conceivably, all of these cases could have been avoided if the systems serving them had effective filtration and adequate disinfection.

The following paragraphs, and corresponding sections of the main document, describe the technologies and costs necessary to achieve adequate control of pathogenic organisms.

TABLE 2. WATERBORNE OUTBREAKS OF GIARDIASIS CLASSIFIED BY TYPE OF WATER TREATMENT OR WATER SYSTEM DEFICIENCY, 1965 TO 1984

Water Source and Treatment/Deficiency	Outbreaks	Cases
1. Surface water source, chlorination only*	39	12,088
2. Surface water source, filtration	15	7,440
3. Surface water source, untreated	12	322
4. Cross-connection	4	2,220
5. Groundwater, untreated:		
a. well water source	4	27
b. spring source	2	44
6. Groundwater, chlorination only		
a. well water source	2	126
b. spring source	2	29
7. Contamination during main repair	2	1,313
8. Contamination of cistern	1	5
9. Consumption of water from nonpotable-tap	1	7
10. Consumption of water while swimming, diving	2	90
11. Insufficient information to classify	4	65
TOTAL	90	23,776

* Includes three outbreaks and 76 cases of illness where filtration was available but not used. In one outbreak filtration facilities were used intermittently and in two outbreaks filtration facilities were bypassed.

Source: Reference 2.

FILTRATION IN COMMUNITY SYSTEMS (SECTION III)

At this time, treatment requirements for filtration and disinfection are the best available means for controlling pathogenic organisms.

This section provides a description of the characteristics and efficiency of water filtration technologies. The section contains supporting data on each technology in the form of laboratory and pilot-plant studies, and case histories regarding full-scale plant operation. The following paragraphs present a summary of the treatment efficiency of each technology, together with a brief description of each technology.

General

Filtration of domestic water supplies is the most widely used technique for removing turbidity and microbial contaminants. The removal of suspended particles occurs by straining through the pores in the filter bed, by adsorption of the particles to the filter grains, by sedimentation of particles while in media pores, coagulation (floc growth) while traveling through the pores, and, in the case of slow-sand filters, by biological mechanisms.

Effectiveness of Filtration for Removal of Microbial Contaminants

Filtration processes provide various levels of microbial contaminant removal. Tables 3 and 4 summarize microbial removal efficiencies determined from field and pilot plant studies completed on a range of filtration processes.[3,4]

Table 3 includes virus removal results by several filtration processes without disinfection. As shown in the table, all of the processes are capable of removing 99 percent of viruses without disinfection.

Giardia lamblia removal data by conventional treatment, direct filtration, diatomaceous earth filtration, and slow-sand filtration are shown in Table 4. Very high levels (>99.9%) of Giardia reduction can be achieved by chemical coagulation followed by settling and filtration, or by direct filtration. The importance of coagulation to achieve high levels of Giardia removal is noted for both processes. Diatomaceous earth filtration is also extremely effective in removing Giardia cysts. Slow-sand filtration which relies on biological as well as physical mechanisms to remove microbial contaminants is especially effective in removing Giardia cysts.

In his review of performance data, Logsdon compared slow-sand filtration, diatomaceous earth filtration, and conventional and direct filtration.[3] Using information from filtration studies at pilot-scale, full-scale, or both, he showed that all of the filtration processes, when properly designed and operated, can reduce the concentration of Giardia cysts by 99 percent or more, if they are treating a source water of suitable quality. Many of the studies also contained Giardia removals of 99.9 percent, agreeing with the values shown in Table 4.

TABLE 3. REMOVAL EFFICIENCIES OF VIRUSES BY WATER TREATMENT PROCESSES

Unit Process	Percent Removal	Operating Parameters	Reference
Slow sand filtration	99.9999	0.2 m/hr, 11-12°C	16*
	99.8	0.2 m/hr†	
	99.8	0.4 m/hr, 6°C	
	91	0.4 m/hr†	
Diatomaceous earth filtration	>99.95	With cationic polymer coat	17**
	††	Cationic polymer into raw water	
Direct filtration	90-99	2-6 gpm/ft^2, 17-19°C	18*
Conventional treatment	99	2-6 gpm/ft^2, 17-19°C	18*

* Pilot-scale studies.
**Laboratory-scale studies.
† No temperature data given.
††No viruses recovered.

TABLE 4. REMOVAL EFFICIENCIES OF GIARDIA LAMBLIA BY WATER TREATMENT PROCESSES

Unit Process	Raw Water Concentration	Percent Removal	Operating Parameters	Reference
Rapid filtration with coagulation, sedimentation	23-1100/L	96.6-99.9	Min. Alum = 10 mg/L Opt. pH = 6.5 Filtration rate = 4.9-9.8 m/hr ($2.0-4.0\ gpm/ft^2$)	5*
Direct filtration with coagulation	20×10^6/L (as slug)	95.9-99.9	Min. alum = 10 mg/L pH range = 5.6-6.8	5*
- No coagulation		48	Filt. rate = 4.9-9.8 m/hr ($2.0-4.0\ gpm/ft^2$) Eff. NTU/inf. NTU = (0.02-0.5)/(0.7-1.9) Eff. poor during ripening	
- With flocculation		95-99	Alum. = 2-5 mg/L Polymer (Magnifloc 572 CR) = 1.2 mg/L Temp. = 5°-18°C Eff. NTU/inf. NTU = 0.05/1.0 Filt. Rate = 4.8 - 18.8 m/hr ($2.0-7.75\ gpm/ft^2$)	6*
- No coagulation		10-70		6*
Diatomaceous earth filtration	1.5×10^5-9.0×10^5L	99-99.99	Filter aid = 20 mg/L body feed	5**
	10^2-10^4/L	>99.9	Filt. rate = 2.4 - 9.8 m/hr ($1.0-4.0\ gpm/ft^2$) Temp. = 5°-13°C Eff. NTU/inf. NTU = (0.13-0.16)/(1.0-2.0)	7**
Slow sand filtration	$50-5 \times 10^3$/L	100	Filt. rate = 0.04 - 0.4 m/hr Temp. 0°, 5°, 17°C (1.0-10 mgad) Eff. NTU/inf. NTU = (3-7)/(4-10)	8**

* Laboratory and pilot-scale studies.
**Laboratory-scale studies.
Source: Reference 4.

Discussion of Most Applicable Technologies

The following methods of filtration are identified as the Most Applicable Technologies and are those most widely used for removal of turbidity and microbial contaminants:

- Conventional treatment
- Direct filtration (gravity and pressure filters)
- Diatomaceous earth filtration
- Slow-sand filtration
- Package plants

Conventional Treatment--
Conventional treatment is the most widely used technology for removing turbidity and microbial contaminants from surface water supplies. Conventional treatment includes the pretreatment steps of chemical coagulation, rapid mixing, flocculation and sedimentation followed by filtration.

Direct Filtration--
The direct filtration process can consist of any one of several different process trains depending upon the application.[9] In its most simple form, the process includes only filters (oftentimes pressure units) preceded by chemical coagulant application and mixing. Raw water must be of seasonally uniform quality with turbidities routinely less than 5 NTU in order to be effectively filtered by an in-line direct filtration system. A second common configuration of the process includes floccculation as pretreatment for the filters, in addition to coagulant application and mixing. Preflocculation results in better performance of certain dual-media filter designs on specific water supplies.

Diatomaceous Earth Filtration--
Diatomaceous earth (DE) filtration, also known as precoat or diatomite filtration, is applicable to direct treatment of surface waters for removal of relatively low levels of turbidity. Diatomite filters consist of a layer of DE about 1/8-inch thick supported on a septum or filter element. The thin precoat layer of DE is subject to cracking and must be supplemented by a continuous-body feed of

diatomite, which is used to maintain the porosity of the filter cake. If no body feed is added, the particles filtered out will build up on the surface of the filter cake and cause rapid increases in headloss. The problems inherent in maintaining a perfect film of DE between filtered and unfiltered water have restricted the use of diatomite filters for municipal purposes, except under favorable conditions.

Slow-Sand Filtration--

Slow-sand filters are similar to single-media rapid-rate filters in some respects, yet they differ in a number of important characteristics. In addition to (1) slower flow rates (by a factor of 50 to 100 versus direct filtration for example), slow-sand filters also: (2) function using biological mechanisms instead of physical-chemical mechanisms, (3) have smaller pores between sand particles, (4) do not require backwashing, (5) have longer run times between cleaning, and (6) require a ripening period at the beginning of each run.

Package Plants--

Package plants are not a separate technology in principle from the preceding technologies. They are, however, different enough in design criteria, operation, and maintenance requirements that they are discussed separately in this document.

The package plant is designed as a factory-assembled, skid-mounted unit generally incorporating a single, or at the most, several tanks. A complete treatment process typically consists of chemical coagulation, flocculation, settling and filtration. Package plants, for purposes of this document, generally can be applied to flows ranging from about 25,000 gpd to approximately 6 mgd.

DISINFECTION IN COMMUNITY SYSTEMS (SECTION IV)

Section IV provides a description of the process characteristics and inactivation efficiencies of disinfection technologies. The section contains detailed criteria and supporting data on each technology in the form of laboratory, pilot-scale, and plant-scale studies. The following paragraphs present a summary of the inactivation efficiency of each technology, together with a brief description of that technology.

General

While the filtration processes described in Section III are intended to physically remove microbial contaminants from water supplies, disinfection is specifically used to inactivate or kill these organisms. Sterilization, or the destruction of all organisms in water, is not considered. Disinfection is most commonly achieved by adding oxidizing chemicals to water, but can also be accomplished by physical methods (applying heat or lignt), by adding metal ions, or by exposure to radioactivity.

Most Applicable Technologies

The following methods of disinfection are identified as the Most Applicable Technologies (not necessarily in order of effectiveness), and are those most widely used for destruction of microbial contaminants:

- Chlorination (chlorine liquid, gas, and hypochlorite)
- Chlorine dioxide
- Chloramination
- Ozonation

The performance of these and other chemical disinfectants can best be described through the use of the C•T product (the product of residual disinfectant, C, in mg/L, and contact time, T, in minutes). A detailed description of the application of the C•T concept to disinfection practice has been presented by Hoff.[10]

The range of concentrations and contact times for different disinfectants to achieve 99 percent inactivation of E. coli, poliovirus, and Giardia cysts are presented in Table 5. As shown by the concentration-time (C•T) products in the table, there is wide variation both in resistance of a specific organism to the different disinfectants, and in the disinfection requirements for different organisms using a single disinfectant. In general, however, the C•T products in the tables show that Giardia cysts are the most resistant to disinfection, followed by viruses, whereas E. coli are the least resistant.

Chlorination--

For purposes of disinfection of municipal supplies, chlorine is applied primarily in two forms: as a gaseous element, or as a solid or liquid chlorine-containing hypochlorite compound. Gaseous chlorine is generally considered the least costly form of chlorine that can be used in large facilities. Chlorine is shipped in cylinders, tank cars, tank trucks, and barges as a liquified gas under pressure. Chlorine confined in a container may exist as a gas, as a liquid, or as a mixture of both. Thus, any consideration of liquid chlorine includes consideration of gaseous chlorine.

Hypochlorite forms (principally calcium or sodium) have been used primarily in small systems (less than 5,000 persons) or in large systems where safety concerns related to handling the gaseous form outweigh economic concerns. Present day commercial, high-test calcium hypochlorite products contain at least 70 percent available chlorine and are usually shipped in tablet or granular forms. Sodium hypochlorite is provided in solution form, containing 12 percent or 15 percent available chlorine.

Chlorine Dioxide--

Chlorine dioxide (ClO_2) is not widely used as a disinfectant in the United States, though its use for this purpose is relatively common in Europe. Chlorine dioxide cannot be transported because of its instability and explosiveness, so it must be generated at the site of application. The most common method for producing ClO_2 is by chlorination of aqueous sodium chlorite ($NaClO_2$), although the use of sodium chlorate is more efficient. For water treatment, chlorine dioxide is only used in aqueous solutions to avoid potential explosions. In terms of available chlorine, chlorine dioxide has more than 2.5 times the oxidizing capacity of chlorine. However, its comparative efficiency as a disinfectant varies with a number of factors.

Chloramination--

In aqueous systems, chlorine reacts with ammonia (NH_3) to form chloramines. Chloramines have many properties different than chlorine. Chloramines are generally less effective than chlorine for inactivating bacteria, viruses, and protozoans at equal dosages and contact times. Conventional practice for

TABLE 5. SUMMARY OF C.T VALUE RANGES FOR 99 PERCENT INACTIVATION OF VARIOUS MICROORGANISMS BY DISINFECTANTS AT 5°C

	Disinfectant			
Microorganism	Free Chlorine pH 6 to 7	Performed Chloramine pH 8 to 9	Chlorine Dioxide pH 6 to 7	Ozone pH 6 to 7
E. coli	0.034-0.05	95-180	0.4-0.75	0.02
Polio 1	1.1-2.5	770-3,700	0.2-6.7	0.1-0.2
Rotavirus	0.01-0.05	3,800-6,500	0.2-2.1	0.006-0.06
G. lamblia cysts	47->150	-	-	0.5-0.6
G. muris cysts	30-630	-	7.2-18.5	1.8-2.0

Source: Reference 10

chloramination in the field is to add ammonia to the water first, and chlorine later.

Ozonation--

Ozone is the most potent and effective germicide used in water treatment. Only free residual chlorine can approximate it in bactericidal power, but ozone is far more effective than chlorine against both viruses and cysts. However, since it is highly reactive, ozone does not provide a long-lasting residual in drinking water. In addition, ozone must be produced electrically on-site as it is needed, and it cannot be stored.

SMALL WATER SYSTEMS (SECTION V)

This section defines the characteristics of small water systems, and identifies and describes filtration, disinfection, and alternative technologies applicable to small systems. In this document, small water systems are defined as those with design capacities less than 1.0 mgd. They may serve either community systems or noncommunity systems, and often have distinctly different characteristics and problems than larger systems.

Several surveys of small systems were performed in connection with this report to determine characteristics such as system supply capacity, treatment design capacity, average day flow requirements, and system storage capacity.[11,12] The results of those two surveys, coupled with data collected informally from other small water systems, lead to the definition of flow characteristics for four flow categories for the purposes of this report, as defined in Section VI.

The processes and facilities used by existing small water systems to treat water supplies vary about as widely as the range of flows they treat.

Treatment Technologies Applicable to Small Systems

Many of the technologies described in Section III, Filtration, and Section IV, Disinfection, are adaptable to smaller systems. Others, because of such factors as operational complexity, safety considerations, equipment size limitations and cost, are not appropriate for small systems.

Filtration technologies that can serve small systems include:

- Package plants
- Slow-sand filters
- Diatomaceous earth filters
- Ultrafiltration
- Cartridge filters

Disinfection technologies which may be appropriate for small systems include:

- Hypochlorination and gaseous chlorination
- Ozonation
- Iodination (noncommunity systems serving transient populations)
- Erosion feed chlorinators
- Ultraviolet radiation (groundwater systems)

Detailed descriptions of each of these technologies are provided in the main document.

Alternatives to Treatment

Under certain circumstances, some small systems may have alternatives available to them that are not practical for larger systems. Specifically, it may be possible for a small system to construct a well to provide a groundwater source as either a supplement to or a replacement for an existing surface water source. Further, some systems may be small enough that purified water vending machines could be used to supply all or a major portion of its water demand.

COST DATA (SECTION VI)

Capital and operating costs for the technologies in this document are based upon updated costs originally presented in several cost documents prepared for EPA.[13,14] Construction cost information originating from those reports was modified and updated by acquisition of recent cost data. Specific details regarding cost calculations for individual processes and process groups are presented in this section and in Appendix B. Description of assumptions and costs for disinfection of groundwater are presented in Appendix A.

All costs were prepared for facilities with average flows as shown in Table 6, selected after a national survey of operating treatment plants.[15] Capacities range from 26,000 gpd for the smallest plant to 1.3 billion gpd for the largest. Construction costs are based on plant capacity flow rates shown in the table, while operation, maintenance, and chemical use costs are based on average flow rates.

Individual descriptions and costs are presented for the following processes:

- Pumping
  - Package Raw Water Pumping
  - Raw Water Pumping
  - In-Plant Pumping
  - Backwash Pumping
  - Package High-Service Pumping
  - Finished Water Pumping
  - Unthickened Chemical Sludge Pumping
  - Thickened Chemical Sludge Pumping
- Chemical Feed
  - Basic Chemical Feed
  - Liquid Alum Feed
  - Polymer Feed
  - Sodium Hydroxide Feed
  - Lime Feed
  - Sulfuric Acid Feed
- Filtration Process Components
  - Rapid Mix
  - Flocculation
  - Rectangular Clarifiers
  - Tube Settling Modules
  - Gravity Filtration
  - Convert Rapid-Sand Filters to Mixed-Media Filters
  - Filter-to-Waste Facilities
  - Capping Sand Filters with Anthracite
  - Slow-Sand Filters
  - Pressure Filtration
  - Contact Basins for Direct Filtration
  - Hydraulic Surface Wash Systems
  - Washwater Surge Basins
  - Automatic Backwashing Filter
  - Clearwell Storage

TABLE 6. PROPOSED AVERAGE PRODUCTION RATE AND PLANT CAPACITIES

Category	Population Range	Average Flow (Q_A), mgd	Surface Water Treatment: Plant Capacity, mgd	Surface Water Treatment: Q_A as % of Plant Capacity
1	25 - 100	0.013	0.026*	50.0
2	101 - 500	0.045	0.068*	66.2
3	501 - 1,000	0.133	0.166*	80.0
4	1,001 - 3,300	0.40	0.500*	80.0
5	3,301 - 10,000	1.30	2.50	52.0
6	10,001 - 25,000	3.25	5.85	55.6
7	25,001 - 50,000	6.75	11.58	58.3
8	50,001 - 75,000	11.50	22.86	50.3
9	75,001 - 100,000	20.00	39.68	50.4
10	100,001 - 500,000	55.50	109.90	50.5
11	500,001 - 1,000,000	205	404	50.7
12	>1,000,000	650	1,275	51.0

*Costs for supply systems in Categories 1-4 include significant supplemental storage volumes. See Section V of main document for further details.

Package Pressure Diatomite Filtration
Pressure Diatomite Filters
Package Ultrafiltration Plants
Package Conventional Complete Treatment
Disinfection Processes
Chlorine Storage and Feed Systems
Chlorine Dioxide Generation and Feed
Ozone Generation, Feed, and Contact Chambers
Ammonia Feed Facilities
Ultraviolet Light Disinfection
Solids Handling Processes
Sludge Holding Tanks
Sludge Dewatering Lagoons
Liquid Sludge Hauling
Gravity Sludge Thickeners
Filter Press
Dewatered Sludge Hauling
Administration, Laboratory and Maintenance Building

A summary of the total costs (in units of cents/1,000 gallons) of all filtration process groups, disinfection facilities for surface waters, and other individual and alternative processes is presented in Table 7.

GROUNDWATER DISINFECTION COSTS (APPENDIX A)

Cost tables are presented in this appendix for the disinfection of groundwater. A summary of total costs from those calculations is shown in Table 8.

SURFACE WATER FILTRATION COST CALCULATIONS (APPENDIX B)

This appendix contains detailed design criteria and example cost calculations for six different filtration process groups, showing the construction and O&M costs of each process in the plant.

COSTS OF OBTAINING AN EXCEPTION TO THE SURFACE WATER FILTRATION RULE (APPENDIX C)

This appendix contains a set of cost tables that provide estimates of potential incremental costs for water utilities who wish to continue to supply unfiltered surface water to their customers. The tables include both estimated construction and O&M costs for upgraded disinfection facilities, and estimated incremental costs for watershed management, sanitary surveys, and intensive monitoring and analysis of turbidity and bacteriological and physical characteristics of source waters.

COSTS FOR PRESENTLY FILTERING SYSTEMS TO IMPROVE THEIR DISINFECTION FACILITIES (APPENDIX D)

Appendix D contains cost estimates for disinfection system improvements necessary to meet several different criteria. These costs are based on both the performance and the actual facilities in a number of operating filtration plants.

COST RANGE INFORMATION REGARDING LAND, PIPING, AND FINISHED WATER PUMPING (APPENDIX E)

This appendix provides supplemental cost information on system components which can be very different in size or other characteristics depending on conditions at specific treatment plant sites. Consequently, these costs can be highly variable and should be used with caution.

TABLE 7. SUMMARY OF TOTAL COSTS

Treatment Processes	Total Cost of Treatment, ¢/1,000 Gallons Size Category[1]											
	1	2	3	4	5	6	7	8	9	10	11	12
	0.026	0.068	0.166	0.50	2.50	5.85	11.59	22.86	39.68	109.9	404	1,275
	0.013	0.045	0.133	0.40	1.30	3.25	6.75	11.50	20.00	55.5	205	650
Filtration[2]												
Complete treatment package plants	944.5	277.4	195.1	113.6	72.8	52.4						
Conventional complete treatment					104.1	70.3	58.6	61.9	53.8	39.3	32.0	31.0
Conventional treatment with automatic backwashing filters					87.9	58.3	50.8	57.6	49.4	41.5		
Direct filtration using pressure filters			322.7	137.2	79.1	48.8	39.2	45.8	36.9	28.2		
Direct filtration using gravity filters preceded by flocculation				150.2	90.5	58.4	46.8	50.5	39.8	28.6	23.6	21.3
Direct filtration using gravity filters and contact basins				131.2	80.9	54.7	44.2	48.0	37.5	26.3	21.4	19.1
Direct filtration using diatomaceous earth	672.9	227.2	134.7	66.6	43.1	43.1	36.1	48.1	41.7	35.4		
Slow-sand filtration	377.8	205.1	133.4	54.7	34.3	28.7	25.3					
Package ultrafiltration plants	455.6	226.8	179.2	138.4								

1. Category values, from top to bottom, are number, design flow (mgd), and average flow (mgd). Population ranges for each category are:

1. 25 - 100	4. 1,001 - 3,300	7. 25,001 - 50,000	10. 100,001 - 500,000
2. 101 - 500	5. 3,301 - 10,000	8. 50,001 - 75,000	11. 500,001 - 1,000,000
3. 501 - 1,000	6. 10,001 - 25,000	9. 75,001 - 100,000	12. >1,000,000

2. Each process group includes chemical addition and individual liquid and solids handling processes required for operation; excluded are raw water pumping, finished water pumping, and disinfection.

TABLE 7 (Continued)

Treatment Processes	Total Cost of Treatment, ¢/1,000 Gallons Size Category[1]											
	1	2	3	4	5	6	7	8	9	10	11	12
	0.026	0.068	0.166	0.50	2.50	5.85	11.59	22.86	39.68	109.9	404	1,275
	0.013	0.045	0.133	0.40	1.30	3.25	6.75	11.50	20.00	55.5	205	650
Disinfection[3]												
Chlorine feed facilities[4]	65.9	23.6	16.2	9.7	4.3	2.8	2.1	1.6	1.3	1.0	0.8	0.7
Ozone generation and feed[5]	109	37.2	27.5	12.7	7.0	4.5	3.4	2.6	2.2	1.7	1.4	1.2
Chlorine dioxide[6]	322	87.7	46.1	16.8	7.0	4.2	2.9	2.2	1.7	1.3	1.0	0.9
Chloramination[7]	163	51.1	23.9	14.4	6.1	3.6	2.6	2.1	1.6	1.3	1.0	0.9
Ultraviolet light	43.2	14.1	8.4	5.4								
Supplemental Processes												
Add polymer feed, 0.3 mg/L	35.3	11.0	8.2	2.9	2.9	1.2	0.7	0.5	0.3	0.2	0.2	0.1
Add polymer feed, 0.5 mg/L	36.4	11.4	8.4	3.0	3.0	1.4	0.8	0.6	0.4	0.3	0.2	0.2
Add alum feed, 10 mg/L	87.1	26.2	16.5	7.5	1.9	1.1	0.9	0.8	0.7	0.6	0.6	0.5
Add sodium hydroxide feed	30.2	10.7	8.4	4.7	2.7	1.9	1.6	1.6	1.5	1.4	1.4	1.4
Add sulfuric acid feed	27.0	9.9	7.6	4.3	1.0	0.6	0.4	0.3	0.3	0.2	0.2	0.2
Capping rapid-sand filters with anthracite coal	0.5	0.3	0.3	0.3	0.3	0.3	0.3	0.3	0.3	0.3		
Converting rapid-sand filters to mixed-media filters	9.4	5.5	3.3	2.1	2.0	1.7	1.6	1.5	1.5	1.5		

3. Disinfection facilities include all required generation, storage, and feed equipment; contact basin and detention facilities are excluded. Design flows for Categories 1-4 are, respectively: 0.026 mgd, 0.068 mgd, 0.166 mgd, and 0.50 mgd.
4. Dose is 5.0 mg/L; includes hypochlorite solution feed for Categories 1-3, chlorine feed and cylinder storage for Categories 4-10, and chlorine feed and on-site storage for Categories 11 and 12.
5. Dose is 1.0 mg/L.
6. Dose is 3.0 mg/L.
7. Doses are chlorine at 3.0 mg/L and ammonia at 1.0 mg/L.

TABLE 7 (Continued)

Treatment Processes	Total Cost of Treatment, ¢/1,000 Gallons Size Category[1]											
	1	2	3	4	5	6	7	8	9	10	11	12
	0.026	0.068	0.166	0.50	2.50	5.85	11.59	22.86	39.68	109.9	404	1,275
	0.013	0.045	0.133	0.40	1.30	3.25	6.75	11.50	20.00	55.5	205	650
Supplemental Processes (cont.)												
Add tube settling modules	2.7	1.6	0.9	0.7	0.6	0.5	0.4	0.4	0.4	0.4	0.4	0.4
Add contact basins to an in-line direct filtration plant	16.1	10.9	6.0	3.6	3.7	2.6	1.8	1.6	1.3	1.0	1.0	1.0
Add rapid mix	91.3	30.1	20.3	7.9	4.0	2.5	2.0	2.0	1.9	1.9	1.8	1.8
Add flocculation	45.2	20.1	13.3	7.7	6.2	3.7	2.3	2.0	1.7	1.3	1.2	1.2
Add clarification	95.5	41.7	36.4	18.3	12.7	11.2	10.7	10.1	9.8	9.5	9.2	9.2
Add hydraulic surface wash	80.1	29.0	13.3	5.5	2.4	1.4	1.1	0.9	0.7	0.7	0.7	0.6
Add filter-to-waste facilities	5.5	4.0	2.4	1.3	0.9	0.4	0.3	0.2	0.2	0.2	0.1	0.1
Finished water pumping[8]	70.3	23.7	12.7	7.0	18.4	16.0	14.9	14.3	13.9	13.7	13.7	13.7
Alternatives to Treatment[9]												
Construct new well, 350 ft	303.0	117.5	59.5	35.0								
Bottled water vending machines	795.5	478.1										

8. Facilities include a package high service pumping station for Categories 1-4, and a custom-designed and constructed station for Categories 5-12.
9. Design flows are equal to system demand, i.e.,

Category	Design Flow, mgd	Average Flow, mgd
1	0.07	0.013
2	0.15	0.045
3	0.34	0.133
4	0.84	0.400

TABLE 8. SUMMARY OF GROUNDWATER DISINFECTION COSTS

DISINFECTION METHOD	COST OF TREATMENT, ¢/1,000 GAL											
	Size, Category											
	1	2	3	4	5	6	7	8	9	10	11	12
	0.06	0.14	0.31	0.96	3.06	7.52	15.4	25.3	44.2	124	465	1,505
	0.013	0.045	0.133	0.40	1.30	3.25	6.75	11.5	20.0	55.5	205	650
Chlorine[1]	66.9[2]	22.9	12.0	4.7	3.1	1.6	1.1	0.9	0.8	0.6	0.4	0.3
	30.2	19.4	8.7	6.2	5.4	3.3	2.5	2.0	1.7	1.3	1.0	0.8
	97.1[2]	42.3	20.7	10.9	8.5	4.9	3.6	2.9	2.5	1.9	1.4	1.1
Ozone[3]	165.1	61.6	31.6	15.5	7.5	5.0	4.0	3.5	3.1	2.1	1.4	1.2
	-	-	-	-	-	-	-	-	-	-	-	-
	165.1	61.6	31.6	15.5	7.5	5.0	4.0	3.5	3.1	2.1	1.4	1.2
Chlorine Dioxide[4]	321.3	96.9	45.3	16.8	6.0	3.1	2.1	1.6	1.4	0.9	0.6	0.4
	19.7	10.0	4.7	3.3	3.7	2.4	1.6	1.3	1.1	0.8	0.5	0.4
	341.0	106.9	50.0	20.1	9.7	5.5	3.7	2.9	2.5	1.7	1.1	0.8
Chloramination[5]	116.8	41.8	23.6	9.5	3.9	1.9	1.2	1.0	0.8	0.6	0.3	0.2
	30.2	19.4	8.7	6.2	5.4	3.3	2.5	2.0	1.7	1.3	1.0	0.8
	147.0	61.2	32.3	15.7	9.3	5.2	3.7	3.0	2.5	1.9	1.3	1.0
Ultraviolet Light[6]	61.9	20.2	10.1	7.6								
	-	-	-	-								
	61.9	20.2	10.1	7.6								

[1] Costs include a chlorine dosage of 2.0 mg/l with 30-minute detention time. Chlorine is fed as a hypochlorite solution for Categories 1 through 4; cylinder storage and feed is used for Categories 5 through 10, with on-site tank storage used in Categories 11 and 12. Detention storage is provided by a pressure vessel in Categories 1 and 2, a looped underground pipeline in Categories 3 and 4, and a chlorine contact basin in Categories 5 through 12.

[2] Costs, including both capital and O&M, are shown as follows:

xx.x - disinfection generation/feed equipment
yy.y - detention storage facilities
zz.z - total cost

[3] Includes direct in-line ozone application at a dosage of 1.0 mg/l followed by a 5-minute contact time, assumed to be achieved in the transmission line between the well and the distribution system.

[4] Includes chlorine dioxide at a dose of 2.0 mg/l with a 15-minute detention time. Detention is in pressure vessels in Categories 1 and 2, in looped underground pipelines in Categories 3 and 4, and in chlorine contact basins in Categories 5 through 12.

[5] Costs are for a chlorine dose of 1.5 mg/l, an ammonia dose of 0.5 mg/l, and 30-minute detention. Chlorine is provided as a hypochlorite solution in Categories 1 through 4, by cylinder storage and feed in Categories 5 through 10, and is stored in on-site tanks in Categories 11 and 12. Ammonia is fed as anhydrous ammonia in Categories 1 through 4, and as aqua ammonia in Categories 5 through 12. Detention is in pressure vessels in Categories 1 and 2, in looped underground pipelines in Categories 3 and 4, and in chlorine contact chambers in Categories 5 through 12.

[6] No additional detention storage used beyond that built into the ultraviolet light unit.

I. Introduction

PURPOSE OF THE DOCUMENT

This document assists the Administrator of the U.S. Environmental Protection Agency (EPA) in identifying the best technologies or other means that are generally available, taking costs into consideration, for inactivating or removing microbial contaminants from surface water and groundwater supplies of drinking water. For municipal officials, engineers, and others, the document provides a review of alternative technologies and their relative efficiency and cost. More specifically, this document discusses water treatment technologies which may be used by community and noncommunity water systems in removing turbidity, *Giardia*, viruses, and bacteria from water supplies. EPA is currently developing treatment regulations addressing these microbiological concerns. While most of this document is devoted to discussion of surface water supplies, a brief discussion of disinfection technologies and costs for removing microbial contaminants from groundwater supplies of drinking water is also provided in Appendix A of this document, since disinfection is the best available technology for groundwater systems to comply with the coliform regulations.

The technologies and actions available to a community searching for the most economical and effective means to comply with the microbiological regulations include modification of existing treatment systems; installation of new treatment systems; selection of alternate raw water sources; regionalization; and documenting the existence of a high quality source water while implementing an effective and reliable disinfection system, combined with a thorough monitoring program, and maintaining a continuing compliance with all drinking water regulations. Potential costs for the latter, nonfiltration alternative are presented in Appendix C.

It is not the intent of EPA to require any system to use a particular technology to achieve compliance with the proposed treatment regulations. Instead, the responsibility is retained by the individual water systems to select one or more procedures that are optimal for their particular water supply situation. Whichever individual or combination of technologies is ultimately selected by a water supplier to achieve compliance with the requirements must be based upon a case-by-case technical evaluation of the system's entire treatment process, and an assessment of the economics involved. However, the major factors that must be considered include:

- Quality and type of raw source water
- Raw water turbidity
- Type and degree of microbial contamination
- Economies of scale and the potential economic impact on the community being served
- Treatment and waste disposal requirements

The information provided herein is intended to aid a system in reviewing available technologies for achieving the required reduction in turbidity and microorganisms. It provides the user with an evaluation of the various methods in use today for the removal of different concentrations of turbidity and microorganisms, as well as relative costs.

Some methods are more complex or more expensive than others. Selection of a technology by a community may require engineering studies and/or pilot-plant operations to determine the level of removal a method will provide for that system.

Alternative technologies for the removal of microbial contaminants and turbidity are identified because of their adaptability to treatment of drinking water supplies. It is expected that after development and pilot-scale testing, these methods may be technically and economically feasible for specific situations.

DEFINITION OF TECHNOLOGY CATEGORIES

The methods that can be applied for the removal of microbial contaminants are divided into three categories:

Most Applicable Technologies: Those that are generally available and have a demonstrated removal or control based on experience and studies for most systems subject to the regulations, and for which reasonable cost estimates can be developed.

Other Applicable Technologies: Those additional methods not identified as generally available, but which may have applicability for some water supply systems in consideration of site-specific conditions, despite their greater complexity and cost.

Additional Technologies: Those experimental or other methods with potential use that may be studied for specific situations to achieve compliance with the regulations, and for which insufficient data exist to fully evaluate the suitability and applicability of the technology for removal of microbial contaminants.

MOST APPLICABLE TECHNOLOGIES

The Most Applicable Technologies are water treatment processes or technologies within the technical and financial capability of most public water systems. Some technologies identified may be more complex and expensive than others but because of site-specific conditions and system size, are more applicable and effective for removing microorganisms of concern. Prior to implementing a technology, site-specific engineering studies of the methods identified to remove microbial contaminants should be made. The engineering study should select a technically feasible and cost-effective method for the specific location where microbial contaminant removal is required. In some cases, a simple survey may suffice, whereas in others, extensive microbiological/chemical analyses, design and performance data will be required. The study may include laboratory tests and/or pilot-plant operations to cover seasonal variations in water quality, preliminary designs and estimated capital and operating costs for full-scale treatment.

The technologies identified as Most Applicable Technologies are considered to be the most widely used methods for achieving compliance with the Maximum Contaminant Level (MCL). The Most Applicable Technologies for meeting turbidity and microbial contaminant removal standards are listed as follows:

Filtration

- Conventional treatment (coagulation, flocculation, sedimentation, and filtration)
- Direct filtration using gravity or pressure filters (with coagulation)
- Diatomaceous earth filtration
- Slow-sand filtration
- Package plants

Package plants are listed and described as a separate treatment technology for several reasons: (1) although they are considered to be miniature conventional plants, they usually contain different unit processes than conventional plants (for example, tube settlers are often used instead of clarifiers); (2) design criteria are often different than for larger conventional plants; (3) operation and maintenance requirements are different from those in larger plants; and (4) their use is restricted to a relatively small range of design flows.

Disinfection

- Chlorination (chlorine liquid, gas, and hypochlorite)
- Chlorine dioxide
- Chloramines
- Ozone

OTHER APPLICABLE TECHNOLOGIES

Technologies classified as Other Applicable Technologies which may have application for removal of turbidity and microbial contaminants are listed as follows:

Filtration

- Ultrafiltration (membrane)
- Cartridge (ceramic tube)

Disinfection

- Iodine
- Bromine
- Ultraviolet light
- Heat treatment

ADDITIONAL TECHNOLOGIES

No Additional Technologies for filtration have been identified. Additional disinfection technologies include:

- Silver (impregnated on granular activated carbon)
- Gamma radiation
- Insoluble ion exchange resins

Filtration technologies are discussed and described in Section III. Disinfection technologies are discussed and described in Section IV.

Section V provides a description of the characteristics of small water systems and treatment technologies applicable to these systems.

Cost estimates developed for the most applicable technologies and other applicable technologies, for various system size categories, are presented in Section VI.

II. Background

GENERAL

It was not until the development of the germ theory of disease by Pasteur in the 1860's and the verification of that theory by Koch in the mid-1880's that disease transmission by water could be understood. The connection between contaminated water and disease had been suggested by Dr. John Snow in London in 1854 during his now famous epidemiological study of the "Broad Street Well" cholera epidemic. He concluded the well had become contaminated by a visitor who arrived in the vicinity with the disease. Cholera was one of the first diseases to be recognized as capable of being waterborne. Cholera and other diseases prevalent in the 19th and early 20th centuries are now effectively controlled by water treatment and hygienic distribution. However, the number of known microorganisms capable of causing waterborne diseases continues to grow, and new concerns about disease transmission, including viruses and parasites, are still being identified.

An overview of trends in the incidence of waterborne disease over the past 40 years, and of the capabilities and limitations of available treatment processes, indicates the need for: (1) constant awareness of a broad spectrum of disease-producing organisms; (2) continuing improvement in microbial detection techniques; and most importantly, (3) proper application and operation of available treatment processes. All of these practices are safeguards to prevent water distribution systems from becoming conduits of disease.

WATERBORNE DISEASE OUTBREAKS - 1946 TO 1980

A thorough review of 35 years of information about waterborne disease outbreaks has been presented by Lippy and Waltrip.[1] Available data from the U.S. Public Health Service (USPHS), the Center for Disease Control (CDC), and the U.S. EPA, were characterized in terms of annual occurrence, type of water system involved, responsible-system deficiencies, causative microbial agent, and geographic distribution.

Annual occurrences of outbreaks of waterborne diseases from 1946 to 1980 are shown on Figure II-1, which indicate a general decline from 1946 to 1966, but a rapid rise from less than 10 outbreaks to 50 thereafter. As stated by the authors, the increase in annual outbreak occurrences is probably due to both more active data collection by federal agencies, and more aggressive investigating and reporting by a few states.[1] The most recent data, therefore, appear more representative of actual outbreaks than data collected prior to 1967, but still lack accuracy due to the absence of intensive surveillance by many state and local agencies.

Statistics regarding the general classes of pathogens and chemical agents which produced the waterborne disease outbreaks and cases of illness are shown in Table II-1. Of the identified causative agents, bacterial pathogens accounted for both the greatest number of outbreaks (146 outbreaks or 45 percent) and the greatest number of illnesses (36,682 cases or 56 percent). Viral contaminants produced 25 percent of the outbreaks, but only 8 percent of the cases of illness. Over the last 10 years, parasitic contaminants have been identified often enough to have produced 15 percent of the total number of outbreaks, and 30 percent of the total cases of illness over the entire 35-year study period.

Of equal importance in Table II-1 is the fact that more than half of both the number of outbreaks and the cases of illness were caused by unidentified factors. This fact indicates the current difficulty with procedures permitting rapid detection of waterborne disease, as well as diagnosis and cause identification. Quite often, a disease outbreak has peaked and subsided before health officials are able to investigate and determine the cause.

Microbial contaminants causing waterborne diseases over the 35-year study period are also indicated in Table II-1. Among the individual causative agents, Giardia ranks number one in cases of illness and number four in outbreaks, even though it was not identified as a causative agent for the first time until the mid-1960's. The bacterial pathogens Salmonella and Shigella rank first and third, respectively, in outbreaks caused, and second and third, respectively, in cases of illness produced. Hepatitis, a viral pathogen caused the second highest number of

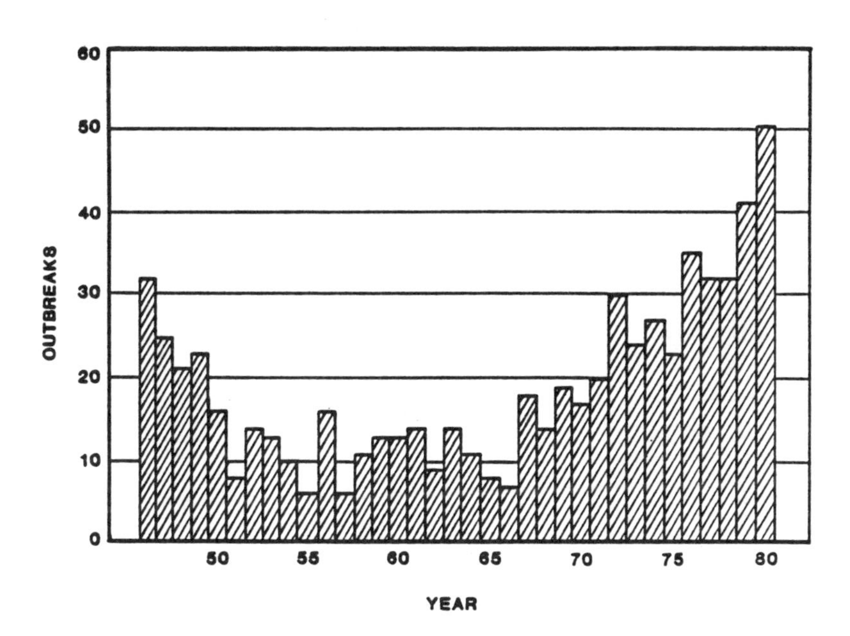

Figure II-1
ANNUAL OCCURRENCE OF WATERBORNE DISEASE OUTBREAKS

TABLE II-1. CAUSATIVE AGENTS OF WATERBORNE DISEASE, 1946 TO 1980

Agent	Outbreaks	Cases of Illness
Bacterial		
Campylobacter	2	3,800
Pasteurella	2	6
Leptospira	1	9
Escherichia coli	5	1,188
Shigella	61	13,089
Salmonella	75	18,590
TOTAL	146	36,682
Viral		
Parvovirus-like	10	3,147
Hepatitis	68	2,262
Polio	1	16
TOTAL	79	5,425
Parasitic (Protozoan)		
Entamoeba	6	79
Giardia	42	19,734
TOTAL	48	19,813
Chemical		
Inorganic	28	891
Organic	21	2,725
TOTAL	49	3,616
Unknown	350	84,939
GRAND TOTAL	672	150,475

Source: Reference 1.

outbreaks, but produced relatively few identifiable cases of illness. A second example of the importance of Giardia and viruses as predominant causative agents is shown in Table II-2.

TABLE II-2. ETIOLOGY OF WATERBORNE OUTBREAKS IN THE UNITED STATES, 1972-1981

	Outbreaks	Cases
Acute gastrointestinal illness	183	37,069
Giardiasis	50	19,863
Chemical poisoning	41	3,717
Shigellosis	22	5,105
Hepatitis A	10	282
Salmonellosis	8	1,150
Viral gastroenteritis	11	4,908
Typhoid fever	4	222
Campylobacter gastroenteritis	4	3,902
Toxigenic E. coli gastroenteritis	1	1,000
Cholera	1	17
TOTAL	335	77,235

Source: Reference 2.

The types of water systems in which disease outbreaks occurred are shown in Table II-3. Definitions of the types of systems listed in Table II-3, taken from present drinking water regulations, are as follows:

- Community System: Serves at least 15 connections or 25 residents on a year-round basis.

- Noncommunity System: Serves at least 15 connections or a daily average of 25 persons, at least 60 days of the year (for water systems serving transient populations).

- Individual System: Any water system not defined as a community system or a noncommunity system.

TABLE II-3. OCCURRENCE OF WATERBORNE DISEASE OUTBREAKS, BY TYPE OF WATER SYSTEM

System Type	Outbreaks 1946 to 1980	Outbreaks 1971 to 1980	Outbreak Occurrence per 1,000 Systems*
Community	237	121	1.9 (121/65)
Noncommunity	296	157	1.0 (157/150)
Individual	137	36	
Unknown	2	0	

*For the 10-year period 1971 through 1980.

Source: Reference 1.

Several factors combine to make the data from community systems more accurate than those from other types of systems. Actual outbreaks which occurred in noncommunity and individual systems are, undoubtedly, significantly higher than shown in Table II-3, simply due to less accurate reporting. However, estimates of total actual cases of illness in community systems still far outnumber those in noncommunity systems. In the most recent years shown, 1971 to 1980, there were between 13 and 14 reported outbreaks per year in community systems.

Water systems deficiencies contributing to or causing disease outbreaks between 1946 and 1980 are ranked in Table II-4. In this table, inadequate or interrupted treatment includes equipment breakdown, insufficient chlorine contact time, and overloaded processes. Distribution problems include contamination from cross-connections, improper or inadequate main disinfection, and open distribution reservoirs. Miscellaneous items consist of intentional, malicious contamination of a water supply, and undetermined causes of outbreaks. Review of the table indicates that more than 80 percent of all outbreaks in public water systems were caused by either use of untreated, contaminated groundwater, or poor practices in treatment and distribution of water. One specific comment by Lippy and Waltrip regarding water system deficiencies is worth noting here:

> Unfortunately, the analyses...show that the causes of outbreaks from 1946 to 1980 do not markedly differ from those presented in summaries for 1920 to 1945, which may indicate that drastic changes are needed in attitudes of regulatory agencies and the industry.[1]

TABLE II-4. RANKING OF WATER SYSTEM DEFICIENCIES CAUSING DISEASE OUTBREAKS IN PUBLIC WATER SYSTEMS, 1946 TO 1980

Rank	Public Water System Deficiency	Percent of Outbreaks
1	Contaminated untreated groundwater	35.3
2	Inadequate or interrupted treatment	27.2
3	Distribution network problems	20.8
4	Contaminated, untreated surface water	8.3
5	Miscellaneous	8.3

Source: Reference 1.

The most glaring deficiencies found at disease outbreak sites were that disinfection was not in place where it was needed, or not properly operated where it was in place.

A more recent (1965-1984) evaluation of system deficiencies specifically responsible for giardiasis outbreaks is presented in Table II-5. It is worth noting that 83.5 percent of the cases listed in Table II-5 (first three items, surface water systems) occurred in systems that either did not have filtration, or in which filtration was ineffective or intermittent. Conceivably, all of these cases could have been avoided if the systems serving them had effective filtration and adequate disinfection.

WATERBORNE DISEASE OUTBREAKS - 1981 TO 1983

Information from annual summaries of the Center for Disease Control for the years 1981, 1982, and 1983 tends to confirm trends which began in the mid-1970's, regarding waterborne diseases caused by microbial contaminants.[3] Neither the number of outbreaks nor the cases of illness declined significantly in these years, and the same water system deficiencies that caused disease outbreaks in the 1920's are still causing illnesses in the 1980's.

TABLE II-5. WATERBORNE OUTBREAKS OF GIARDIASIS CLASSIFIED BY TYPE OF WATER TREATMENT OR WATER SYSTEM DEFICIENCY, 1965 TO 1984

Water Source and Treatment/Deficiency	Outbreaks	Cases
1. Surface water source, chlorination only*	39	12,088
2. Surface water source, filtration	15	7,440
3. Surface water source, untreated	12	322
4. Cross-connection	4	2,220
5. Groundwater, untreated:		
a. well water source	4	27
b. spring source	2	44
6. Groundwater, chlorination only		
a. well water source	2	126
b. spring source	2	29
7. Contamination during main repair	2	1,313
8. Contamination of cistern	1	5
9. Consumption of water from nonpotable-tap	1	7
10. Consumption of water while swimming, diving	2	90
11. Insufficient information to classify	4	65
TOTAL	90	23,776

* Includes three outbreaks and 76 cases of illness where filtration was available but not used. In one outbreak filtration facilities were used intermittently and in two outbreaks filtration facilities were bypassed.

Source: Reference 2.

Disease outbreaks for each of the years, 1981 to 1983, and the type of water system involved, are shown in Table II-6. The total number of outbreaks declined from the peak of 50 reached in 1980 (Figure II-1), to the range of 32 to 40 annual occurrences found also from 1976 through 1979. The cases of illness dropped dramatically from 1980 (20,008 cases) to 1981 and 1982 levels, but returned to the 1980 level in 1983 (20,905 cases). Of the total cases of illness in 1983, 11,400 occurred during one outbreak, of undetermined cause, involving persons attending a religious festival at a campground in Pennsylvania. The trend for reporting more outbreaks and illnesses occurring in community systems rather than noncommunity systems started in the late 1970's. That trend has continued, possibly implying closer surveillance and better reporting from systems with a resident population.

TABLE II-6. WATERBORNE DISEASE OUTBREAKS BY TYPE OF SYSTEM, 1981 TO 1983

	Community		Noncommunity		Private		Total	
Year	Outbreaks	Cases	Outbreaks	Cases	Outbreaks	Cases	Outbreaks	Cases
1981	14	3,104	16	1,322	2	4	32	4,430
1982	22	2,028	12	1,330	6	98	40	3,456
1983	29	8,845	6	11,875	5	185	40	20,905

Source: Reference 3.

The etiology and type of water system involved in 1981 to 1983 disease outbreaks are displayed in Table II-7. Unlike the 1946 to 1980 period, the percent of outbreaks with unknown causes (represented by acute gastrointestinal illness of unknown origin (AGI) in CDC data) was always less than 50 percent in 1981, 1982, and 1983, indicating improvement in detection, diagnosis, and reporting practices. Of the identified causative microbial agents, the parasite _Giardia_ accounted for 55 to 71 percent of outbreaks during the 3-year period. _Giardia_ was the most frequently identified pathogen for all 3 years and has been so since 1978, although its increased isolation rate can be primarily attributed to more active investigation of unfiltered water systems in Colorado. The outbreaks caused by bacterial pathogens and by viral pathogens were approximately equal over the same 3-year period.[4,5]

Water system deficiencies responsible for 1981 to 1983 waterborne diseases are identified in Table II-8. As noted earlier, the general and specific types of deficiencies are basically unchanged from the 1946 to 1980 period, and remain the same as those cited as far back as the 1920's. Treatment deficiencies accounted for 39 percent of the cases in the three recent years, up from about 32 percent in earlier years. Use of contaminated, untreated groundwater was involved in 29 percent of 1981 to 1983 outbreaks, down from 35 percent in the preceding 35 years. Together with use of contaminated, untreated surface water, however, treatment deficiencies and lack of treatment continue consistently to be responsible for approximately 80 percent of all outbreaks of waterborne disease due to microbial contamination.

TABLE II-7. WATERBORNE DISEASE OUTBREAKS, BY ETIOLOGY AND TYPE OF SYSTEM

Agent	Community			Noncommunity			Private Water Systems			Total Outbreaks		
	1981	1982	1983	1981	1982	1983	1981	1982	1983	1981	1982	1983
AGI*	2	6	8	12	9	6	0	1	1	14	16	15
Giardia	8	9	16	1	2	0	0	1	1	9	12	17
Hepatitis A	0	2	1	0	0	0	0	1	2	0	3	3
Norwalk	0	4	0	0	0	0	0	0	0	0	4	0
Salmonella	0	0	2	0	0	0	0	0	0	0	0	2
Shigella	0	1	0	1	1	0	0	0	1	1	2	1
Yersinia	0	0	0	0	0	0	0	1	0	0	1	0
Campylobacter	1	0	1	0	0	0	0	0	0	1	0	1
V. Cholerae	0	0	0	1	0	0	0	0	0	1	0	0
Rotavirus	1	0	0	0	0	0	0	0	0	1	0	0
Chemical	2	0	1	1	0	0	2	2	0	5	2	1
TOTALS	14	22	29	16	12	6	2	6	5	32	40	40

* AGI = Acute gastrointestinal illness of unknown etiology.

Source: Reference 3.

TABLE II-8. WATERBORNE DISEASE OUTBREAKS, BY TYPE OF DEFICIENCY, 1981 TO 1983

Deficiency	Total Outbreaks 1981	1982	1983	3-Year Period
Untreated surface water	4	5	3	12
Untreated groundwater	9	11	12	32
Treatment deficiencies	11	14	19	44
Deficiencies in distribution system	2	5	4	11
Miscellaneous	3	3	0	6
Multiple deficiencies	3	2	2	7
TOTAL	32	40	40	112

Source: Reference 3.

PROBLEMS IN REMOVING MICROBIAL CONTAMINATION FROM WATER SUPPLY SOURCES

The material just presented on the frequency and severity of waterborne disease outbreaks due to transmission of microbial contaminants through public water supplies roughly defines the minimum extent of this public health problem as it is known today. Due to nonrecognition and incomplete reporting of outbreaks of waterborne disease, the number of actual occurrences is believed to be much higher than that presently reported.[1]

With proper application and use of today's water treatment technology, disease outbreaks from public water supplies could virtually be eliminated. There are numerous reasons why this is not presently the case. The statistics on reported outbreaks of waterborne disease strikingly demonstrate the principal problem which currently exists. Water systems either do not have the needed water treatment facilities, the existing water treatment facilities are inadequate, or there has been an interruption in the proper operation or use of existing facilities.

Interruptions in proper operation can be due to lack of adequate system reliability, mechanical failure, unanticipated emergency conditions, or operator error. Within these general problem categories there are a host of detailed problems which contribute to the present lack of total effectiveness in removing microbes from water. Some of these problems will be presented in the discussions which follow.

False Sense of Security

Public water supplies in the United States have long had the reputation of being excellent with respect to quality and safety. Many protected surface water supply sources have operated successfully in the past with only disinfection treatment. Many do not have backup disinfection equipment for use in times of emergency. To a great extent, the fact that there has never been a recognized problem, explains the present lack of more extensive treatment facilities. Many managers of water systems are not convinced of the need for other than minimum treatment (disinfection) of their supplies, because they have not recognized any problems. Times have changed, however, and the known hazards of contamination have significantly increased in recent years.

Protected Watersheds--

There are several changes affecting the need for treatment of surface water supplies. Many protected watersheds and raw water reservoirs are now opened to the threat of human pollution by the widespread use of trail bikes, four-wheel drive vehicles, and snowmobiles which enable people to reach formerly inaccessible areas. However, this hazard of increased opportunity for human transmission of microbial contaminants, is overshadowed by the potential for transmission of *Giardia* cysts from beaver and other animals to man through drinking water. Pluntze, in describing the first waterborne outbreak of giardiasis having strong evidence of animal origin (Camas, Washington, 1976) aptly stated: "This changed forever the time-honored concept of the 'protected' watershed: i.e., if one could simply keep people out, or at least monitor and control their activities, there would be little likelihood of human waste contaminating the water."[6] The cause of the disease in Camas is presumed to be beavers, not people. "*Giardia* organisms can and do infect many domestic and wild animals, including cats, dogs, sheep, mice, rats, gerbils, beavers, muskrats, and voles."[7]

On protected watersheds, loggers and recreationists are more easily controlled than beavers and muskrats, a situation which is causing some reassessment concerning the safety of unfiltered surface water supplies, even those that have been derived from watersheds which were considered to have maximum protection from human pathogens.

Turbidity Standard--
The NIPDWR (National Interim Primary Drinking Water Regulations) turbidity standards of 1.0 NTU (nephelometric turbidity units) has been used for many years to support and justify nontreatment of raw waters of high clarity. The degree of risk involved in this practice is now recognized as being higher than previously estimated, since Giardia and other cysts have been found in unfiltered waters with turbidities less than 1.0 NTU. Several examples of this type of occurrence are described in the following pages.

Monitoring Microbial Contaminants

Microbial agents that cause waterborne disease outbreaks are rarely isolated from the water system.[8] Examination of water samples for pathogenic bacteria, viruses, and protozoa is technologically and economically infeasible for many water systems. For example, since one plaque-forming viral unit may be an infective dose for a portion of the population, isolating and detecting viruses is quite difficult.[9]

Coliforms as Indicator Organisms--
The fact that the number of pathogens in water is low, and that they occur in wide variety (requiring different methodologies for detection), imposes a severe restriction on their direct and quantitative determination in routine water analysis. As a result, it is necessary to resort to indirect evidence of their presence. Indicator organisms provide the substitute evidence. The presence of coliforms in any water sample indicates the potential for recent fecal contamination, which in turn suggests the possible presence of pathogens. In treated water, their relative numbers are primarily used to indicate treatment efficiency.

Investigation for the presence of coliform organisms, because of the relatively large numbers present in most waters, offers a practical approach for assessing treatment efficiency and detecting contamination. A search for specific pathogenic organisms is likely to take longer than coliform determinations, and might be fruitless.[10,11] Moreover, throughout the longer time required to make and confirm the necessary examinations for specific pathogens, people would continue to consume the water. This delay in determining the hazardous nature of a water could easily result in a waterborne-disease outbreak.

Although the waterworks profession relies on the coliform examination to determine both treatment efficiency and the probability level that disease-producing microorganisms are present, this test also has its limitations. For example, specific pathogens respond differently to the various treatment processes. Also, it may not be correct to assume that treatment reduces all pathogenic organisms to the same extent that it reduces coliform bacteria.

Accumulated evidence indicates that the bacteria causing typhoid, paratyphoid, cholera, and bacillary dysentery do respond to treatment in the same manner as coliform bacteria. Some viruses and cysts, however, appear to persist in water for longer times than coliforms. Viruses may penetrate through rapid-sand filters more readily than coliform bacteria. Certain viruses and cysts are more resistant than coliforms to destruction by chlorine disinfection. Recent work reported by Rose, et al, suggests that enteric viruses can occur at detectable levels in filtered finished water which meets current coliform (1/100 mL) and turbidity (1 NTU) standards and contains >0.2 mg/L free chlorine.[12] From these observations, it can be concluded that it is possible for certain pathogenic organisms to survive treatment that apparently removes or destroys all coliform bacteria.

Despite this possibility, the record of waterborne-disease outbreaks attributable to properly treated public water supplies, as indicated by coliform absence, supports the use of coliforms as one indicator of the microbiological safety of water. With good filtration and disinfection practices there is an excellent chance of removing or inactivating any virus or Giardia that may be present in the raw water.

The limitations of using coliform absence as an indicator of treatment effectiveness are, however, growing more apparent. The AWWA Committee on the Status of Waterborne Diseases in the U.S. and Canada (1981) stated the problem as follows:[4]

> Coliform organism identification is used as an indication of fecal contamination of water supplies and is widely employed for routine surveillance. Negative results are usually interpreted as assurance that the water is free of enteric pathogens. This interpretation must be reevaluated, as outbreaks of waterborne disease have occurred in water systems where coliforms have either not been detected or have not been found to exceed standards.

The methods of sampling for coliforms are limitations in themselves, i.e., coliform samples are not taken continuously, and often very few samples are taken.

Finally, investigations into outbreaks in Camas, Washington; Berlin, New Hampshire; and six different outbreaks in Colorado, from 1976 through 1982, indicate that coliform counts are poor or inadequate indicators regarding Giardia, since treated water concentrations did not violate bacterial standards, yet outbreaks occurred.[13,14,15]

Several different tests for coliforms have been used, including total coliform, fecal coliform, heterotrophic plate count (HFC), and others. Regulations to date have conventionally used total coliform counts, and in some cases, fecal coliforms. The use and value of HPC has also been noted by several authors and groups. McCabe et al, for example, note that this general bacterial enumeration does not usually have direct health significance, but heavy growths do indicate a potential for contamination.[45] They also cite research findings suggesting that high plate counts inhibit the growth of coliforms on lab media, thereby observing their presence.[45] Geldreich et al, later confirmed this finding, and refined it by noting that the frequency of detection of both total coliform and fecal coliform began decreasing after the HPC exceeded about 500/ml.(B) They also found increased difficulty in screening for Salmonella and Shigella serotypes when the HPC exceeded 500/ml. The survey data of Geldreich et al, also included disinfection control results in the distribution system, finding that (a) HPC in distribution lines was controlled to less than 500/ml by maintaining a chlorine residual of 0.3 mg/L, and further, that (b) HPC values less than 10/ml were obtained in more than 60 percent of the distribution systems with chlorine residuals of 0.1 to 0.3 mg/L. Based on these data, the authors recommended establishing an HPC limit of 500/ml in distribution systems.[46] More recently, the AWWA Committee on Heterotrophic Plate County bacteria has recommended that (a) systems monitor for HPC in finished waters at a frequency equal to 10 percent of that for total coliform measurement in the distribution system (minimum of 2 samples per month), and that (b) systems provide treatment which achieve HPC concentrations of less than 10/ml.[47]

Turbidity as an Indicator of Water Quality

Measurement of the turbidity of treated water serves several purposes. Turbidity is a direct indicator of water clarity. Turbidity removal also affords one of the best tests available for the rapid evaluation of the efficiency of the chemical coagulation and filtration processes in removing particulate matter. In conventional treatment, low turbidity water and effective turbidity removal are necessary to ensure proper disinfection. Particles causing turbidity may interfere with the disinfection processes by coating, adsorbing, or otherwise shielding the microorganisms from contact with the disinfectant. Experience in the operation of water systems has long ago established the relationship between low finished water turbidities and improved public health statistics. However, recent experience with the increased incidence of giardiasis focuses even more sharply on the necessity for maintaining very low finished water turbidities in conventional treatment processes to curtail the spread of this disease, although there is some recent data to indicate that slow-sand filters achieve good cyst removal with finished water turbidities greater than 1 NTU.[16,17]

As noted by the 1981 AWWA Committee on the Status of Waterborne Diseases:[4]

> Turbidity that interferes with disinfection is health related; however, forms of turbidity such as iron precipitate or other inorganic matter are not health related. Waterborne disease outbreaks, primarily giardiasis, have occurred in systems where the health-related turbidity limit has not been exceeded. It cannot be assumed that meeting the turbidity limit will prevent waterborne disease when the raw water source may contain pathogens, especially *Giardia* cysts. Under these conditions, safe drinking water can be assured only by properly

designed and operated water filtration plants utilizing coagulants or filter aids in addition to disinfection.

A principal conclusion of the investigation of the Camas, Washington, giardiasis outbreak was that, "....turbidity and coliform count alone are inadequate parameters on which to judge the biological quality of filter effluent."[13,15]

Karlin and Hopkins made the following conclusions regarding Colorado giardiasis outbreaks:[14]

- Turbidity and bacterial concentrations are poor indicators regarding *Giardia*--no violations of standards were associated with these outbreaks.

- A turbidity standard of 1 NTU permits many Colorado water treatment systems to discontinue pretreatment when raw water source is <1 NTU, a frequent occurrence, when *Giardia* in significant concentrations have been found at turbidities as low at 0.1 NTU in clear mountain streams.

In earlier work regarding virus, Robeck and his coworkers concluded, "When a turbid water was treated, a floc breakthrough was usually accompanied by an increase in virus penetration even though the finished water turbidity remained below 0.5 NTU...."[18]

Similarly, during pilot plant work on cyst removal by granular media filtration, Logsdon, et al, concluded that:[19]

- Meeting the 1-NTU MCL for turbidity did not result in effective cyst removal.

- Increases in filtered water turbidity, as low as 0.05 to 0.10 NTU, were usually associated with large increases in cyst concentrations.

Furthermore, in work with asbestos fiber removal which is relevant to removal of *Giardia* cysts from water, Logsdon reported:[20]

Even though turbidity cannot directly measure asbestos fibers in the concentrations found at water treatment plants, when a granular media filtration plant is properly operated, turbidity readings can be used as a guide to plant operation. Filtered water turbidity should be 0.10 NTU or lower to maximize fiber removal. Turbidity increases of 0.1 to 0.2 NTU above this value generally were accompanied by large increases in asbestos fiber concentrations.

Particle Counting--

By use of recently developed sophisticated equipment for particle counting, it is possible to obtain a more precise description of particles in water than is afforded by conventional turbidity measurements.[21] With instrumentation currently available, the number of particles within given size ranges can be measured. From these data, particle size distribution and particle volume can be estimated. Such information can then be analyzed to more thoroughly evaluate treatment process performance.

Particle counting is especially useful in research and pilot-plant work, but also has potential for improved plant operational control. A total particle counter provides a direct measurement of the particulate matter present in water, and in contrast to turbidity measurements, is not influenced by particle size, shape, or refractive index. For these reasons, the total particle counter offers promise of being a very useful and sensitive process control tool in water treatment practice, especially when asbestiform fibers are present in water supplies.[22] However, existing particle counters are designed to be accurate for particles with at least one dimension greater than 1 µm, and in some waters, submicron particles may predominate. In these cases, both turbidimetric and particle size distribution measurements would be necessary.[23]

Treatment for Control of Bacteria and Viruses

In every treatment process the final barrier for control of bacteria and viruses is disinfection. In the case of surface water treatment, it is usually necessary to pretreat the water with a minimum of coagulation and filtration to remove algae, turbidity, iron, manganese, organics, or other interfering substances prior to final disinfection. However, in the past, some surface water sources which are exceptionally clear (turbidity under 5 NTU) have used chlorination as the only treatment provided.

The necessity to remove interfering substances prior to disinfection, and the definite relationship between drinking water clarity, disinfection, and public health is presented by Hudson, as summarized below:[24]

- Filtration plants operated to attain a high degree of removal of one impurity tend to accomplish high removals of other suspended materials. Examples of parallelism in removal of turbidity, manganese, microorganisms, and bacteria are cited.

- Speed and simplicity make the turbidity measurement a valuable index of removal of other materials. Plants producing very clear water also tend to secure low bacterial counts accompanied by low incidence of viral disease.

- The production of high-quality water requires striving toward high goals as measured by several--not just one or two--quality criteria. These criteria include filtered-water turbidity, bacteria as indicated by plate counts and by presumptive- and confirmed-coliform determinations, and thorough chlorination.

The current status of virus removal by treatment can be described by the combined work of several investigators. Leong, through an extensive literature review, found that complete water treatment (coagulation, flocculation, sedimentation, filtration, and disinfection) can remove at least eight logs of virus under optimal conditions.[25] Gerba later tabulated (Table II-9) the substantial reductions of enteric viruses achievable by water treatment processes involving flocculation, filtration, and chlorination.[26] As shown in Table II-9, filtration without pretreatment cannot be depended upon for efficient virus removal. Despite the effectiveness of combined processes in the laboratory, however, field studies described in more detail later in this report show that proper operation is also necessary to achieve high levels of virus removal.[27,28,29] Gerba has also pointed out that: (1) much of the available information about virus removal comes

TABLE II-9. ESTIMATED VIRUS REMOVALS BY WATER TREATMENT PROCESSES FROM LABORATORY STUDIES

Process	Percent Removal
Coagulation/Flocculation/Sedimentation	
Ferric chloride	99.5
Lime	98.8
Alum	95
Filtration (without pretreatment)	73
Disinfection	
Free chlorine	$\geq$99.9
Ozone	$\geq$99.9

Modified from Leong (1983).
Source: Reference 26.

from laboratory studies using a limited variety of laboratory-grown viruses and coliphages, and (2) few studies have been done on the removal of naturally occurring enteric viruses by actual water treatment plants.[26]

Chlorination is also highly effective for inactivation of viruses, and its effectiveness increases as turbidity decreases (fewer protective particles present). For maximum virucidal effect, an AWWA committee has recommended a turbidity less than 1 NTU, a free chlorine residual of 1.0 mg/L or greater for 30 minutes, and a pH of less than 8.0.[30] At the present time, there are no specific standards for the concentration of free chlorine in finished water, only a recommendation that 0.2 mg/L be maintained. One subsection of the NIPDWR (141.21(h)) also allows reduced coliform monitoring procedures if a free chlorine residual above 0.2 mg/L is maintained in the distribution system.

Three recent studies of virus removal at operating water treatment plants indicate that a wide range of removal efficiencies is possible using the same treatment processes, and that operation and operating conditions are critical to effective virus removal. Investigations by Stetler and Gerba resulted in the values for percent virus removals shown in Table II-10.[28,29] While the plants evaluated by Stetler had an average enterovirus removal rate of greater than 98 percent in finished water, the average for those visited by Gerba et al was far lower (81 percent).[28,29] The low average removal found by Gerba et al, however, was significantly affected by poor operating conditions at one plant.

TABLE II-10. VIRUS REMOVAL AT TWO WATER TREATMENT PLANTS

Parameter	Reference 28 Cumulative Percent Removal			Reference 29 Cumulative Percent Removal		
	Clarified	Filtered	Finished	Clarified	Filtered	Finished
Enterovirus	71	90	>98	36	62	81
Rotavirus	ND	ND	ND	24	0*	81

ND = Not Determined
*Value is correct; probably explained by viral water quality variability.
Source: Reference 26.

By-Products of Disinfection--
In the treatment of drinking water, certain disinfectants may react with organic materials present in surface waters to form undesirable organic by-products. Since about 1974 much attention has been given to the formation of chloroform and other trihalomethanes in the chlorination process, because these products are carcinogens. On November 29, 1979, EPA promulgated an amendment to the NIPDWR establishing an MCL for total trihalomethanes of 0.10 mg/L, as well as monitoring and reporting requirements for trihalomethanes (THM's).

When using chlorine dioxide as a water disinfectant, a concern exists about the acute toxicity of chlorine dioxide, chlorite, and chlorate for those people who are exceptionally sensitive to agents producing hemolytic anemia.[31] Chlorine dioxide also possesses antithyroid activity.[31] Because of these concerns, the EPA recommends that total residual levels of these reaction products of chlorine dioxide should be less than 1.0 mg/L in any part of the distribution system.

Chloramines may cause a potentially serious problem when present in tap water used in hemodialysis in artificial kidney machines. Chloramines pass through the dialysis membrane and their toxicity to patients under dialysis conditions is well defined.[32] Operators of dialysis centers are made aware that tap water must be treated before use in dialysis, but cases of illness due to chloramines, chlorine dioxide, or other chemicals continue to be reported.[33] The Association for the Advancement of Medical Instrumentation has proposed a limit of 0.1 mg/L for chloramines in hemodialysis water, in response to the cumulative evidence described above.[34] Monochloramine possesses mutagenic properties in bacteria that are commonly associated with carcinogens.[24]

Presently there are no known harmful by-products from ozonation, but there is a lack of information on the organic reaction products of ozone. For example, some organic materials not completely oxidized by ozone could be more harmful than the original organics in the raw water. However, ozone does not supply a long-lasting germicidal residual when used alone, so it should be used with a disinfectant with a residual to address distribution system concerns.[43]

Treatment for Control of Giardia and Removal of Turbidity

While stringent disinfection alone may be adequate for some well-protected raw water supplies that are both exceptionally clear and free of dissolved interfering substances, it is not sufficient by itself for turbid raw waters and is also insufficient for some seemingly pure, exceptionally clear waters. Case histories and recent evaluations of successes and failures of treatment processes, described in the following paragraphs, lead to the conclusion that filtration with coagulation and disinfection, in most cases, are necessary to control Giardia. In all situations, vigilant operation and maintenance are necessary to ensure continuous protection against cysts and parasites like Giardia.

Removal of turbidity to low levels is desirable both for aesthetic and public health reasons. For most filtration processes, i.e., conventional treatment, direct filtration, and package plants, filtration supplemented by pretreatment processes is necessary to remove turbidity to well below 1.0 NTU in order to provide adequate protection against Giardia. If raw water turbidities are low (1 to 5 NTU), it is essential that high percent turbidity reductions be achieved across the filter to ensure that filtration is effective.

Case Histories--

Information about outbreaks of giardiasis in the mid-1970's has been provided by a number of investigators, and can be summarized as follows:

- Rome, New York, experienced an outbreak of an estimated 4,800 to 5,300 cases of giardiasis in 1974 and 1975. At the time of the outbreak, Rome treated its raw surface water source solely by chlorine and ammonia, producing a chloramine with a combined chlorine residual of only 0.8 mg/l. In addition, higher than normal coliform counts were present prior to the outbreak.[35]

- In 1976, Camas, Washington, had an outbreak of 600 cases of giardiasis out of a population of 6,000. Its surface water source was treated by prechlorination in the transmission line to the plant, addition of a coagulant, and filtration through multimedia pressure filters. However, a number of deficiencies were found in the condition and operation of the filter and in chemical pretreatment, including failure to feed coagulants to water being filtered. *Giardia* cysts were isolated in the surface water entering the plant, and *Giardia*-positive beavers were trapped in the watershed.[36]

- In 1977, 750 cases of giardiasis were reported in Berlin, New Hampshire, due to the community's filtered water supply. Raw water was obtained from two river sources from which *Giardia* cysts were recovered. Water from one source received pressure filtration without chemical pretreatment, at a facility that was in poor condition. Water from the other source was treated by a new plant that provided chemical addition (alum, polymer, and sodium hydroxide), upflow clarification, and rapid sand filtration. Faulty construction at the new plant, of a common wall between the raw and treated water, allowed about 3 percent of the filter influent to bypass the filters. Both plants provided postchlorination.[37]

Of the numerous investigations conducted from 1980 to 1985 regarding outbreaks of waterborne giardiasis, the Colorado study described by Karlin and Hopkins is representative of the range of conditions found in nearly all other studies.[14] Specifically, six outbreaks of giardiasis were confirmed within a 2-year period. In two of those outbreaks, chlorination was the only treatment. In three other water supply systems, filtration and chlorination, without pretreatment, were used. The sixth system used treatment consisting of prechlorination, sedimentation, dual-media filtration, and chlorination (no coagulation or flocculation). In all four treatment systems using more than chlorination, construction or operating faults existed which led to poor cyst removal.

Both of the above investigators concluded that the practice of providing only chlorination, for high-quality surface waters from protected watersheds, is not adequate. Water system reliability requires more than one barrier against disease transmission.

Lin reviewed 24 outbreaks of waterborne giardiasis occurring between 1971 and 1978, and found that ten were in systems where a surface water source was treated with chlorine only, and eight were in systems with untreated surface water.[38] Of the remaining six systems, three had ineffective filtration, two used untreated groundwater supplies, and the cause of one outbreak was undetermined. He also stated that there is no evidence that outbreaks of waterborne giardiasis have ever occurred in systems with well-constructed, well-operated, multiple-process treatment plants.

Argument for Multiple Barriers--
The difficulty in removing Giardia cysts from water by any single treatment process has given added support to the "multiple barrier" concept in water treatment process selection. The multiple barrier approach adds greatly to the safety of water production and the overall reliability of the chain of treatment processes. This is a conclusion reached by the AWWA Committee on the Status of Waterborne Diseases in the U.S. and Canada in 1984, as summarized in the quotation which follows:[4]

> Simple disinfection as the only treatment for surface water sources is ineffective in preventing waterborne transmission of giardiasis. All surface water should receive pretreatment and filtration in addition

> to disinfection. Both pressure and gravity sand filters have proven ineffective in removing *Giardia* cysts under conditions of poor, or simply casual, operation. This has occurred primarily in systems where the raw water turbidity was low. Under these conditions, turbidity removal has been achieved without coagulants but passes *Giardia* cysts. Outbreak data, engineering experience, and filtration theory indicate that *Giardia* cysts can be reduced dramatically by properly functioning conventional sand filters, but the water must be effectively pretreated before filtration. Effective pretreatment includes coagulation, flocculation, and settling prior to filtration, or, if the settling process is not used, the addition of appropriate chemicals for conditioning the water or filter media. State drinking water regulations adopted by Colorado in 1977 require the filtration of surface water sources to remove *Giardia* cysts.

Additional support for the multiple barrier treatment concept is given by numerous water supply professionals. According to Logsdon and Lippy, for example:[5]

- The multiple barrier concept has evolved in water treatment, based on the concept that when one barrier fails, the remaining barriers reduce the impact of the failure.

- Even with advances in water treatment technology, outbreaks of waterborne disease continue to occur because the multiple barrier concept is not being enforced. In many specific cases, filtration is not provided or not properly applied.

ALTERNATIVE APPROACHES TO MICROBIAL CONTAMINANT CONTROL

In the U.S. there is a real opportunity to reduce the occurrence of waterborne disease and improve public health by providing better removal of microbial contaminants from drinking water supplies. Several approaches to bring this about are worthy of consideration.

Quality Requirements

One approach establishes limits for the concentrations of several indicators in water to be delivered to the public through a distribution system. As discussed earlier in this section, the use of indicators presents a number of practical problems, yet at the same time, has been and will continue to be valuable in controlling waterborne disease. For example, the public health benefits of maintaining a finished water turbidity of 0.1 NTU in accordance with the water quality goals of the AWWA were described as long as 17 years ago. In the December 1968 issue of the **American Water Works Association Journal**, is a Statement of Policy by the Association on the "Quality Goals for Potable Water" as reported by the Task Group of AWWA on Water Quality Goals.[39] The rationale for the adopted goal for turbidity is as follows:

> Today's consumer expects a sparkling, clear water. The goal of less than 0.1 NTU insures satisfaction in this respect. There is evidence that freedom from disease organisms is associated with freedom from turbidity, and that complete freedom from taste and odor requires no less than such clarity. Improved technology in the modern treatment processes make this a completely practical goal.

In a modern filter plant which is properly designed and properly operated, finished water turbidities less than 0.10 NTU can be readily maintained in product water. Numerous plants in the U.S. conform to this goal (an AWWA treatment goal) routinely and consistently.[40,41] Most, if not all, of these plants which produce very high quality water have several features in common: (1) they have good control of coagulation either on the basis of frequent jar tests or more commonly by use of continuous automatic monitoring in a coagulant control center; (2) they have facilities to add a filter aid such as a non-ionic polymer in small quantities (0.01 to 0.1 mg/L) to control the depth to which floc penetrates into the filter beds; (3) they have installed dual- or tri-media filters equipped with adequate surface wash and backwash equipment with continuous turbidity monitors on each filter effluent; (4) they have adequate disinfection capability with backup equipment and continuous residual chlorine monitoring; (5) they have adequate instrumentation, control, and standby facilities to provide a high degree of process reliability; and, most importantly, (6) they employ trained and experienced operations and maintenance personnel.

Turbidity goals such as those discussed above appear essential in controlling microbial contamination, but as noted earlier, are not sufficient by themselves to indicate the absence of pathogens like Giardia or viruses.

Surface Water Treatment Requirements

The second approach to providing better removal of microbial contaminants uses mandatory treatment requirements in addition to treated water quality goals, based on performance experience both in research and in public water systems.

The need for filtration of surface water supplies has been suggested based on the recent history of giardiasis outbreaks and the newly acquired knowledge that Giardia cysts can be satisfactorily removed by adequate filtration followed by proper disinfection.[6] In many midwestern and southern states, filtration of all surface waters has been a long-standing requirement of state health agencies. Beavers, muskrats, and other carriers of Giardia cysts appear to pose a significant hazard in water systems which do not filter their supplies.

Providing for filtration of surface water supplies also can eliminate the water quality problems posed by occasional high raw water turbidities and unusual or emergency conditions. The use of filtration to control seasonal turbidity peaks in raw water is preferable to other methods, such as adding untreated well water to a surface water supply in order to reduce turbidities simply by blending. While this practice does have the desired effect on turbidity levels, it also raises the potential risk of adding contaminated water to the distribution system, and does nothing for the removal of Giardia from the raw water.

Conditions When Filtration May Not be Necessary--
Under restricted conditions it may be possible for a water utility to disinfect a surface water supply, without filtration, and still provide the desired level of health protection for its customers.

A description of the basis for such a decision has been presented by Amirtharajah.[44] The article notes that the Safe Drinking Water Act provides both a framework and general principles for variances from a mandatory filtration and disinfection treatment requirement. The principles stated in the Act are based on: (1) the nature of the raw water source, and (2) the use of alternative treatment techniques.

In discussing the nature of the raw water source, Amirtharajah cites criteria which include: (1) an active and thorough watershed control program involving annual on-site sanitary surveys and monitoring for Giardia infested mammals; (2) a consistently high quality water source as documented by a thorough monitoring program (potentially including fecal coliforms, heterotrophic plate count, and turbidity); and (3) stringent and reliable disinfection practices including continuous residual monitoring, complete redundancy of all components (except contact basins), and disinfectant residuals and contact times which will provide safety against Giardia and viruses.[44] Relative safety against these two microbial contaminants is also documented in the article in terms of removal and inactivation percentage, respectively, provided by both filtration processes and disinfectants. Data is presented in the article, similar to Table II-9, showing that filtration plus disinfection can provide 99.9 percent removal of Giardia, and 99.99 percent (or greater) removal of virus. To provide equal health protection, therefore, systems with disinfection only must be able to provide equal reductions in Giardia and virus concentrations, or reductions to equal finished water concentration. Amirtharajah presents data showing that very stringent disinfection conditions (high disinfectant residuals or long contact times) will be necessary to achieve equal protection against Giardia.[44] As an example, the recommendations of the AWWA Committee on Viruses of 1.0 mg/l free chlorine residual for 30 minutes at a pH less than 8, would inactivate less than 99 percent of Giardia cysts, when compared to data presented by Amirtharajah, rather than the 99.9 percent shown to be possible with filtration and disinfection together.

Another approach which proposes a case-by-case quantitative method of determining whether or not filtration may be necessary has been described by Regli et al.[42] Based on the degree of acceptable health risk (to a utility or regulatory agency), the method can provide an estimate of treatment required, and the size of sampling program necessary to demonstrate that the desired level of health protection is provided. The authors identify Giardia cysts and viruses as the organisms of principal interest, and present removal and inactivation rates for a variety of treatment processes and disinfectants. The authors also note that the proposed method is limited in its present usefulness by: (1) a lack of both analytical detection methods and inactivation and removal rate data, and (2) potential cost of the extensive monitoring programs likely to be required.

III. Filtration in Community Systems

GENERAL

Filtration of domestic water supplies is the most widely used technique for removing turbidity and microbial contaminants. Filtration is generally provided by passing water through a bed of sand or a layer of diatomaceous earth, or through a combination of coarse anthracite coal overlaying finer sand. Filters are classified and named in a number of ways. For example, based on application rate, sand filters can be classified as either slow or rapid, yet these two types of filters differ in many more characteristics than just application rate, such as removal process, bed material, method of cleaning, and operation. Based on the type of bed material, filters can be classified as sand, diatomaceous earth, dual-media (coal-sand), or even tri-media in which a third sand layer is added. As indicated above, however, classification by a single method usually implies many other differences.

The removal of suspended particles occurs by straining through the pores in the filter bed, by adsorption of the particles to the filter grains, by sedimentation of particles while in media pores, coagulation (floc growth) while traveling through the pores, and, in the case of slow-sand filters, by biological mechanisms. In water treatment practice where rapid rate granular filters are used, coagulation of colloidal material and larger particles is necessary in order to remove and retain the particulate material in the filter bed. Cleaning is accomplished by backwashing the filter beds to dislodge the accumulated material and flushing it to waste.

EFFECTIVENESS OF FILTRATION FOR REMOVAL OF MICROBIAL CONTAMINANTS

Filtration processes provide various levels of microbial contaminant removal. Tables III-1 and III-2 summarize microbial contaminant removal results of studies completed on a range of filtration processes.

TABLE III-1. REMOVAL EFFICIENCIES OF VIRUSES BY WATER TREATMENT PROCESSES

Unit Process	Percent Removal	Operating Parameters	Reference
Slow sand filtration	99.9999	0.2 m/hr, 11-12°C	3*
	99.8	0.2 m/hr†	
	99.8	0.4 m/hr, 6°C	
	91	0.4 m/hr†	
Diatomaceous earth filtration	>99.95	With cationic polymer coat	25**
	††	Cationic polymer into raw water	
Direct filtration	90-99	2-6 gpm/ft , 17-19°C	36*
Conventional treatment	>99	2-6 gpm/ft^2, 17-19°C	36*

* Pilot-scale studies.
**Laboratory-scale studies.
† No temperature data given.
††No viruses recovered.
Source: Reference 9

Virus removal achieved by filtration processes are shown in Table III-1. These results, together with others discussed in this section, indicate that filtration without disinfection can remove 99 percent of viruses in water supplies. Giardia lamblia removal data by conventional treatment, direct filtration, diatomaceous earth filtration, and slow-sand filtration are shown in Table III-2. Very high levels (>99.9%) of Giardia reduction can be achieved by chemical coagulation followed by settling and filtration, or by direct filtration. The importance of coagulation to achieve high levels of Giardia removal is noted for both processes. Diatomaceous earth filtration is also extremely effective in removing Giardia cysts. Slow-sand filtration which relies on biological as well as physical mechanisms to remove microbial contaminants is especially effective in removing Giardia cysts.

TABLE III-2. REMOVAL EFFICIENCIES OF *GIARDIA LAMBLIA* BY WATER TREATMENT PROCESSES

Unit Process	Raw Water Concentration	Percent Removal	Operating Parameters	Reference
Rapid filtration with coagulation, sedimentation	23-1100/L	96.6-99.9	Min. Alum = 10 mg/L Opt. pH = 6.5 Filtration rate = 4.9-9.8 m/hr ($2.0-4.0\ gpm/ft^2$)	5*
Direct filtration with coagulation	~20 x 10^6/L (as slug)	95.9-99.9	Min. alum = 10 mg/L pH range = 5.6-6.8	5*
- No coagulation		~48	Filt. rate = 4.9-9.8 m/hr ($2.0-4.0\ gpm/ft^2$) Eff. NTU/inf. NTU = (0.02-0.5)/(0.7-1.9) Eff. poor during ripening	
- With flocculation		95-99	Alum. = 2-5 mg/L Polymer (Magnifloc 572 CR) = 1.2 mg/L Temp. = 5°-18°C Eff. NTU/inf. NTU = 0.05/1.0 Filt. Rate = 4.8 - 18.8 m/hr ($2.0-7.75\ gpm/ft^2$)	6*
- No coagulation		10-70		6*
Diatomaceous earth filtration	1.5 x 10^5 - 9.0 x 10^5/L	99-99.99	Filter aid = 20 mg/L body feed	5**
	10^2-10^4/L	>99.9	Filt. rate = 2.4 - 9.8 m/hr ($1.0-4.0\ gpm/ft^2$) Temp. = 5°-13°C Eff. NTU/inf. NTU = (0.13-0.16)/(1.0-2.0)	7**
Slow sand filtration	50-5 x 10^3/L	~100	Filt. rate = 0.04 - 0.4 m/hr Temp. 0°, 5°, 17°C (1.0-10 mgad) Eff. NTU/inf. NTU = (3-7)/(4-10)	8**

* Studies included laboratory and pilot-scale work.
**Studies were laboratory scale.
Source: Reference 1

In a later review of performance data, Logsdon compared slow-sand filtration, diatomaceous earth filtration, and conventional and direct filtration.[9] Using information from filtration studies at pilot-scale, full-scale, or both, he showed that all of the filtration processes, when properly designed and operated, can reduce the concentration of Giardia cysts by 99 percent or more, if they are treating a source water of suitable quality. Many of the studies also contained Giardia removals of 99.9 percent, agreeing with the values shown in Table III-2.

DISCUSSION OF MOST APPLICABLE TECHNOLOGIES

The following methods of filtration are identified as the Most Applicable Technologies and are those most widely used for removal of turbidity and microbial contaminants:

- Conventional treatment
- Direct filtration (gravity and pressure filters)
- Diatomaceous earth filtration
- Slow-sand filtration
- Package plants

Conventional Treatment

Process Description--

Conventional treatment is the most widely used technology for removing turbidity and microbial contaminants from surface water supplies. Conventional treatment includes the pretreatment steps of chemical coagulation, rapid mixing, flocculation and sedimentation followed by filtration. Disinfection is not included in the flow sheet because it is discussed separately in the next section of this document. The filters can be either sand, dual-media, or tri-media. Site-specific conditions will therefore influence the design criteria for each component of a conventional treatment system. For the purposes of this document, it is assumed that the processes described here are used for raw waters with relatively low turbidity, since the need for treatment is more obvious with highly turbid waters.

Figure III-1 is a flow sheet for a conventional treatment plant. Typically, upon entering the plant, raw water is coagulated with aluminum sulfate (alum), ferric or ferrous sulfate, ferric chloride, or an organic cationic or anionic coagulant. Following addition of coagulants, the flow is subjected to rapid mixing to provide complete dispersion of the coagulant into the raw water. Depending on the type of rapid mixing device, detention times ranging from 30 seconds to 2 minutes are typically provided.

Following flash mixing, the water enters a baffled flocculation basin where the degree of mixing is controlled to produce a readily settleable floc. Typically, mechanically mixed flocculation basins are designed with detention times ranging from 20 to 45 minutes.

Flocculated water is then introduced into a sedimentation basin designed with 1 hour to 4 hours of detention time to permit the flocculated water to clarify. Sedimentation basin overflow rates range from 500 gpd/ft^2 to 1,400 gpd/ft^2 depending on site-specific conditions. These basins are usually designed with mechanical sludge collectors for continuous removal of settled solids. In some applications where sludge accumulates slowly due to low raw water turbidity, the basins may be cleaned manually by draining the basin and hosing the collected sludge to the plant sewer.

A well designed and operated sedimentation basin should provide a high level of turbidity removal. Effluent from the sedimentation basins, to be treated by rapid-sand filters, should have a turbidity of less than 2 NTU. If dual- or tri-media filters are used, applied water turbidities can be as high as 10 NTU.

Water treatment plants using rapid-sand filters are generally designed with a filtration rate of 2 gpm/ft^2. Rapid sand filter media varies in effective size from 0.35 mm to 1.0 mm, with a uniformity coefficient of 1.3 to 1.7. Newer plants using dual- or tri-media filters have a design filtration rate of 4 to 6 gpm/ft^2. Properly operated and using adequate coagulant dosages, a plant designed around rapid-sand filters is generally capable of producing a low turbidity filtered water approaching a value of 0.1 to 0.2 NTU. Plants using dual- or tri-media filters can generally produce a lower filtered effluent turbidity since polymer

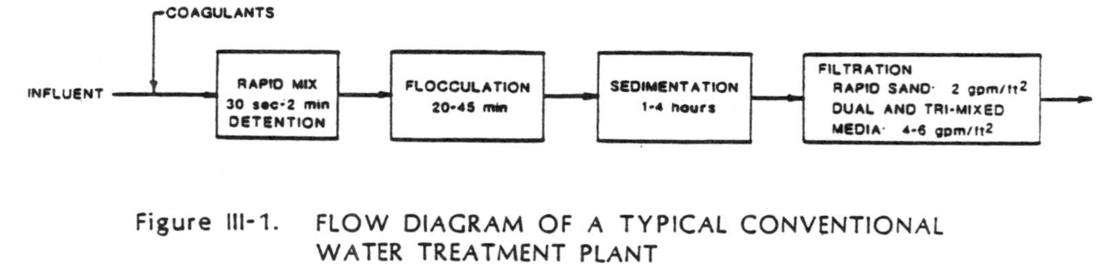

Figure III-1. FLOW DIAGRAM OF A TYPICAL CONVENTIONAL WATER TREATMENT PLANT

filtration aids are generally used to improve filtration performance. Polymers are used in multi-media filters as filtration aids to counteract the higher sheer forces and depth of penetration caused by the higher application rates in dual- or tri-media filters. Polymers are not generally effective as filter aids for rapid sand filters.

Cleasby et al, have noted that since Giardia cysts are relatively large (8 to 14 μm long, 7 to 10 μm wide), properly operated and maintained deep-bed granular media filters should remove them efficiently, when used in a conventional water treatment system.[10,11]

Laboratory and Pilot Plant Studies--

Logsdon et al, have conducted pilot studies showing that removal of Giardia cysts by sedimentation varied from 65 to 93 percent and generally was similar to turbidity removal.[12] The authors also concluded that coarse anthracite media was less efficient at removing Giardia than granular activated carbon (GAC), sand, or dual-media, at similar bed depths. Conventional treatment with any of the three media, however, provided better than 99.9 percent removal of Giardia with optimum chemical dosages, as shown in Table III-3. In his review of filtration studies, Lin concluded that granular media filtration processes with pretreatment are capable of removing more than 99 percent of influent cysts.[13] He also noted that proper coagulation is a necessary key to effective filtration.

A pilot study was undertaken by the University of Washington to evaluate the removal of Giardia lamblia cysts and cyst-sized particles (7 to 12 μm) by drinking water plants.[5] The first phase of the study was devoted to a laboratory-scale evaluation of Giardia removal efficiency by coagulation, flocculation, and mixed-media filtration. The third phase involved on-site testing at Leavenworth and Hoquiam, Washington, with a 20 gpm pilot plant employing coagulation, flocculation, settling, and filtration.

Granular media filtration tests yielded greater than 99.9 percent removal of spiked cysts under optimum conditions. Both the pilot unit and the field unit established the importance of a minimum alum dosage (10 mg/L), an optimum pH range, and intermediate flow rates of 2 to 4 gpm/ft^2. Effluent turbidity and cyst-sized particles passing the filter increased rapidly when the above conditions were not attained or when sudden changes occurred in plant operation.

TABLE III-3. REMOVAL OF G. MURIS CYSTS DURING TEST SERIES 3*

Characteristics of Filter						Turbidity, ntu		Cysts/L		
		Headloss		Rate						
Media	Condition	ft	m	gpm/ sq ft	mm/s	Raw Water	Filtered Water	Applied†	Fil-tered Water	Cyst Removal Percent
GAC	Ripened	4.6-5.2	1.40-1.58	2.42	1.64	8.0-9.5	0.06-0.08	31 000	17	99.94
GAC	Ripening after wash	0.3	0.09	3.04	2.06	7.7	0.17-0.08	31 000	42	99.86
GAC	Backwashed, ripened	0.6-1.4	0.18-0.43	3.04	2.06	7.5-8.5	0.06-0.09	31 000	13	99.958
Sand	Ripening after wash	1.3	0.40	2.86	1.94	8.1	0.14-0.13	31 000	8.3	99.973
Sand	Backwashed, ripened	1.8-5.6	0.55-1.71	2.86	1.94	7.5-8.5	0.07-0.09	31 000	5.2	99.983
Anthracite	Ripened	3.4-4.0	1.04-1.22	2.92	1.98	8.0-9.5	0.10-0.14	31 000	19	99.94
Anthracite	Ripening after wash	0.2	0.06	2.90	1.97	7.7	0.35-0.13	31 000	35	99.89
Anthracite	Backwashed, ripened	0.3-0.5	0.09-0.15	2.90	1.97	7.5-8.5	0.10-0.16	31 000	11	99.96
Dual media	Ripened	2.0-2.8	0.61-0.85	2.90	1.97	7.5-9.5	0.06-0.09	31 000	12	99.96

*pH settled water - 7.2-7.4; alum dosage - 24.8 mg/L; polymer dosage - 0.095 mg/L.
†Cysts dosed to raw water continuously; concentration calculated.
Source: Reference 12

University of Washington conventional-system pilot plant data on cyst removal and mobile pilot plant data on removal of cyst-sized particles suggested that the highest removal percentages for cysts and for 8- to 10-μm particles usually tended to be associated with production of low turbidity water.[5] Best particle removal results were seen for filtered water with turbidities of 0.05 NTU or lower. Below 0.05 NTU effluent turbidity, the median (50 percent of the values) particle removal was 95.1 percent. Between 0.05 and 0.1 NTU effluent turbidity, the median removal was 94.3 percent. At turbidities between 0.5 and 1.0 NTU, the median removal dropped to less than 80 percent. These results demonstrated the close correlation of filtered water turbidity with residual Giardia cysts and underscores the need to produce filtered water of very low turbidity to achieve maximum removal of these organisms.

Case Histories--

Montreal--Seven conventional water treatment plants in the Montreal, Quebec, area that ranged in capacity from 2.4 mgd to 26.4 mgd were sampled twice a month to evaluate the removal of indicator bacteria and cytopathogenic enteric viruses.[4] Samples of raw water, chlorinated raw water, sedimentation basin effluent, filtered effluent, ozonated and finished (tap) water were collected and analyzed. Raw water quality at all seven plants would be regarded as poor, with total coliform counts exceeding 10^5 to 10^6 organisms per liter (10^4 to 10^5 organisms/100 ml), and having an average virus count of 3.3 most probable number of cytopathogenic units per liter (MPNCU/liter); several samples contained more than 100 MPNCU/liter. All plants produced finished water that was essentially free of indicator bacteria as judged by analysis of 1-liter samples for total coliforms, fecal coliforms, fecal streptococci, coagulase-positive staphylococci, and Pseudomonas aeruginosa. The total plate counts at 20 and 35°C (1 mL samples) were also evaluated as an indicator for the removal of the bacteria population and averaged between 10^2 and 10^4 CFU/liter in the finished water. Viruses were detected in 7 percent (11 of 155) of the finished water samples (1,000 liters) at an average density of 0.0006 MPNCU/liter, the highest virus density measured being 0.02 MPNCU/liter. The average cumulative virus reduction was 95.15 percent after sedimentation, and 99.97 percent after filtration (which included some systems using prechlorination), and did not noticeably decrease after ozonation or final chlorination. The viruses isolated from treated waters were all enteroviruses: poliovirus types 1, 2, and 3; coxsackievirus types B3, B4, and B5; echovirus type 7; and untyped picornaviruses.

Denver--The Denver Water Board's 180 mgd Moffat Treatment Plant (Colorado) typically will produce a filtered water turbidity approaching 0.1 NTU with applied water (filter influent) turbidities ranging from 2 to 4 NTU.[14] The filters are preceded by a 4-hour sedimentation basin which provides about 50 percent turbidity removal. The plant has 24-inch deep rapid-sand filters containing 0.50 to 0.55 mm silica sand which have a design rate of 2 gpm/ft^2. Raw water is coagulated with alum, and a nonionic polymer is added as a settling aid. Carry-through of residual coagulant aid also improves turbidity removal by the filters.

Sacramento--The City of Sacramento's American River Plant (California) has a design flow of 105 mgd. It is a conventional water treatment plant consisting of rapid mixing, flocculation, sedimentation, and filtration.[15] Raw water which ranges in turbidity from 2 to 40 NTU, is coagulated with alum at dosages ranging from 15 to 25 mg/L. Flocculation and sedimentation generally produces a settled water turbidity in the 1 to 3 NTU range. Filtration through 24-inch deep tri-media filters routinely produces a high clarity filtered water with turbidities in the 0.08 to 0.09 NTU range. Filter run times of 48 to 72 hours are experienced at filtration rates of 3.5 to 4.5 gpm/ft^2. Prior to the replacement of rapid-sand filter media with tri-media, filtered effluent turbidities averaged 0.2 to 0.4 NTU and filter runs seldom exceeded 24 hours. Subsequent to media replacement, average filtered turbidities have been less than 0.1 NTU. In addition, from July 1985 through June 1986, total coliform counts in finished waters were always less than 2.2/100 ml, and the average total coliform removal for that 12-month period was greater than 99.85 percent.

Direct Filtration

Process Description--

The direct filtration process can consist of any one of several different process trains depending upon the application.[16] In its most simple form, the process includes only dual- or mixed-media filters (oftentimes pressure units) preceded by chemical coagulant application and mixing. The mixing requirement, particularly in pressure filters, can be satisfied by influent pipeline turbulence. In larger plants with gravity filters, an open rapid-mix basin with mechanical mixers is typically used. Raw water must be of seasonally uniform quality with turbidities routinely less than 5 NTU in order to be cost-effectively filtered by a system (in-line) using a flow sheet such as illustrated in Figure III-2. Cost-effective in this context means producing a treated water of a required quality at acceptable operating costs.

Cleasby et al, have noted that direct in-line filtration produces a relatively poor quality filtrate at the beginning of filter runs and therefore requires a filter-to-waste period.[10,11]

Another variation of the direct filtration process is shown in Figure III-3. The direct filtration arrangement shown in Figure III-3 consists of the addition of a coagulant to the raw water followed by rapid mixing and flocculation. The chemically conditioned and flocculated water is then applied directly to a dual- or tri-mixed media filter. Preflocculation results in better performance of certain dual-media filter designs on specific water supplies. Investigators have noted in some cases that preflocculation improves turbidity removal but can lead to premature floc breakthrough.[17] However, incorporating preflocculation does not permit higher turbidity loadings to the filters.

Figure III-4 is a flow sheet of a direct filtration process utilizing a 1-hour contact basin between the rapid mix basin and the filter. The contact basin is designed to promote mixing and contact of chemical floc with turbidity in the water. It also serves as a silt and sand trap. Prechlorination contact time for control of microbial contaminants is also provided by this basin. The contact basin increases plant reliability by adding lead time to smooth-out sudden variations in raw water turbidity. Typically, the basins do not have sludge collectors and must be manually cleaned. Direct filtration plants featuring contact basins have been used extensively and successfully in the Pacific Northwest and in Colorado.

The principal attraction of direct filtration over conventional treatment processes is the potential savings in total cost that can amount to as much as 30 percent each where direct filtration is applicable.

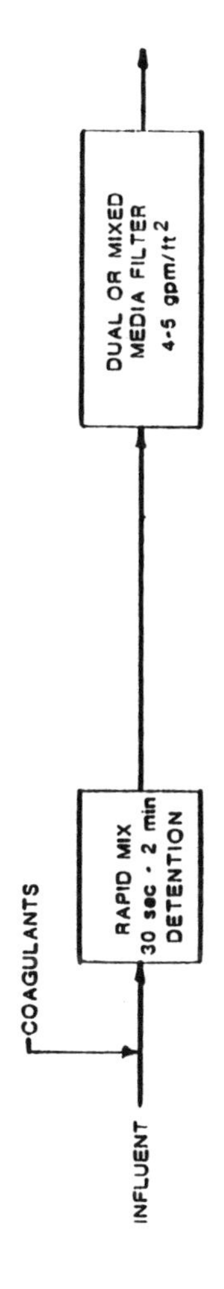

Figure III-2. FLOW DIAGRAM FOR A TYPICAL DIRECT FILTRATION PLANT

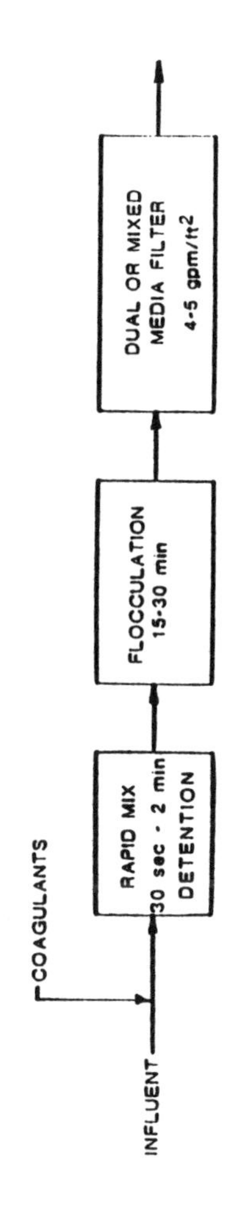

Figure III-3. FLOW DIAGRAM FOR A TYPICAL DIRECT FILTRATION PLANT WITH FLOCCULATION

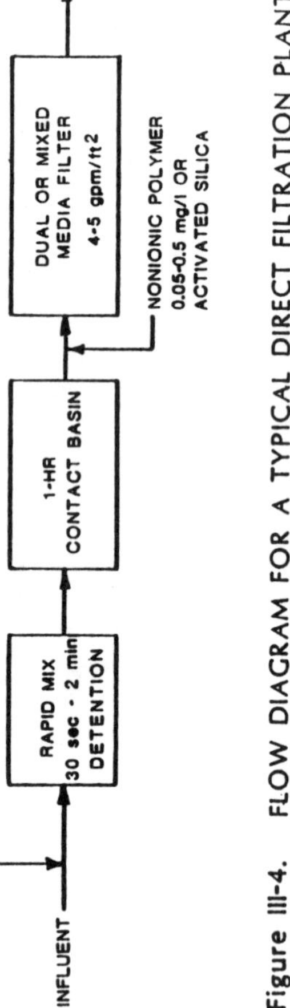

Figure III-4. FLOW DIAGRAM FOR A TYPICAL DIRECT FILTRATION PLANT WITH A CONTACT BASIN

Direct filtration can be used where the raw water is of satisfactory quality. As reported by the Direct Filtration Subcommittee of the AWWA Filtration Committee, the possibilities of applying direct filtration to municipal plants are good if certain requirements can be met.[17] In general, waters with less than 40 units of color, turbidity consistently below 5 NTU, iron and manganese concentrations less than 0.3 mg/L and 0.05 mg/L, respectively, and algae counts less than 2,000 ASU/mL (aereal standard units) appear to be good candidates for direct filtration. Bacteria, viruses, and Giardia are effectively removed if coagulants are applied, effective turbidity and color removal are consistently attained, and efficient chlorination is practiced.

The suitability of direct filtration cannot be determined from numerical values of raw water quality alone; such values only provide a preliminary indication. Pilot plant tests must be performed in each case to verify whether or not direct filtration will provide satisfactory treatment under the site-specific conditions.

Direct filtration without presedimentation using pressure filters is possible where raw water turbidities are generally less than 5 NTU. Over a dozen direct filtration plants ranging in capacity from less than 0.2 mgd to 44 mgd are in operation in Northern California. Raw water is obtained from lakes, reservoirs, and directly from rivers. In general, raw water turbidities average less than 1 to 2 NTU to peak values of 20 to 50 NTU during storms in the winter, when water demands are much less than summer requirements. Treatment consists of prechlorination, coagulation with cationic polymers, filtration through dual media or trimedia filters, and disinfection. The design filtration rate for most facilities is 5 gpm/ft^2. Coagulant dosages range from 0.5 to 2 mg/L and typically produce filtered effluent turbidities of 0.2 to 0.3 NTU. Because source water turbidities are generally low, filter runs between backwashes range from 24 to 96 hours.

Laboratory and Pilot Plant Studies--

Pilot plant work at Colorado State University investigated the removal of Giardia cysts by direct filtration for a range of operating conditions, using water having turbidity levels below 1 NTU and temperatures ranging from 0° to 17°C.[18,6] The most important finding of this work is that proper chemical pretreatment is imperative if direct filtration is to be effective when using low turbidity waters. Specific conclusions of the work included the following:

- With no chemical pretreatment, removal of Giardia, bacteria, and turbidity can be expected to vary between 0 and 50 percent.

- Improvement in removal efficiency was not significant when ineffective coagulants or improper dosages were used.

- With proper chemical pretreatment, removal of all constituents can be expected to exceed 70 percent for turbidity, 99 percent for bacteria, and 95 percent for Giardia cysts.

- Filtration with effective coagulation (i.e., adequate to reduce turbidity from about 0.5 NTU to 0.1 NTU) was capable of removing 95 to 99.9 percent of Giardia cysts and 95 to 99.9 percent of total coliform bacteria.

The roles of other process variables were not as important as chemical pretreatment. Direct filtration without flocculation (Figure III-2) was as effective as direct filtration with flocculation (Figure III-3). Single medium (sand) and dual media (anthracite and sand) both have the same efficiencies in reducing turbidity and bacteria. Hydraulic loading rate has very little effect on removals in the range between 1 and 8 gpm/ft^2. At 10 gpm/ft^2, a moderate negative impact on removals occurs. Investigation of temperature influence showed no trend in removals of turbidity, bacteria, and Giardia cysts at 5°C compared with removals at 18°C.

One phase of the pilot plant work by the University of Washington investigated the effectiveness of direct filtration to remove Giardia cysts from drinking water.[5] At optimum conditions, cyst removal was consistently high. For example, an alum dosage of 12 mg/L, pH 6.2, and a filter loading rate of 2 gpm/ft^2, resulted in 99.73 percent cyst removal at the end of a 1-hour filter ripening period, and 99.94 percent cyst removal later in the same run when effluent turbidity was constant at 0.02 NTU. Reduction in coagulant dose led to an increase in the number of cysts passing through, as did raising the pH to 6.8.

An examination of Table III-4 shows that filtered water turbidity could be used as a guide to cyst removal efficiency in direct filtration.[5] Cyst removal

TABLE III-4. CYST REMOVAL DURING DIRECT FILTRATION AT UNIVERSITY OF WASHINGTON PILOT PLANT

Run No.	Alum Coagul. and Dosage (mg/L)	pH	Filter Loading Rate (m/hr)	Cyst Addition: Total Dosage	Filter Influent Dosage	Cyst Removal (%)	Elapsed Time (Hr:Min)	Influent Turbidity (ntu)	Effluent Turbidity (ntu)	Turbidity Removal (%)
72	None	6.5	6.1	$2.0 \cdot 10^6$	$6.6 \cdot 10^5$	48	4:30	0.73	0.39	47
73	12.0	6.2	6.1	$12.7 \cdot 10^6$	$3.8 \cdot 10^6$	99.73	1:15	1.24	0.03	98
			4.3	$20.0 \cdot 10^6$	$4.2 \cdot 10^6$	99.943	26:00	1.19	0.19	84
74	12.0	6.2	9.7	$15.5 \cdot 10^6$	$7.3 \cdot 10^6$	99.936	1:00	1.37	0.04	97
76	12.0	6.2	9.4	$19.0 \cdot 10^6$	$8.7 \cdot 10^6$	99.979	7:00	1.14	0.02	98
77	7.0	6.2	9.7	$19.2 \cdot 10^6$	$9.8 \cdot 10^6$	99.75	1:00	1.94	0.24	88
			8.6	$20.4 \cdot 10^6$	$9.4 \cdot 10^6$	99.87	16:00	0.81	0.03	96
78	4.0	6.2	10.1	$21.4 \cdot 10^6$	$10.7 \cdot 10^6$	64	2:30	1.31	0.52	60
			8.6	$20.2 \cdot 10^6$	$8.8 \cdot 10^6$	91.8	72:30	1.35	0.37	73
79	12.0	6.8	9.7	$21.5 \cdot 10^6$	$10.3 \cdot 10^6$	95.4	1:00	0.95	0.28	71
			8.3	$21.5 \cdot 10^6$	$8.4 \cdot 10^6$	99.41	10:00	1.02	0.04	96
80	12.0	5.6	9.7	$20.4 \cdot 10^6$	$10.0 \cdot 10^6$	99.83	1:00	1.73	0.03	98
81	12.0	5.6	9.7	$20.4 \cdot 10^6$	$9.8 \cdot 10^6$	99.84	7:00	1.78	0.02	99
	CatFloc		9.7	$20.1 \cdot 10^6$	$9.8 \cdot 10^6$	95.9	1:00	0.92	0.23	75
82	T-1	6.4	9.7	$20.1 \cdot 10^6$	$9.8 \cdot 10^6$	99.911	21:00	0.80	0.27	66
	5.0									

Source: Reference 5

exceeded 99.9 percent ten times out of 15 runs. In seven of the ten instances, the filtered water turbidity was less than 0.10 NTU. Cyst removal was less than 99 percent on five occasions, and each time the filtered water turbidity exceeded 0.10 NTU (actual range was 0.19 to 0.52 NTU). When no coagulants were used during filtration, the process removed only 48 percent of the spiked cysts and 47 percent of the turbidity.

Case Histories--

Lake Oswego--The Lake Oswego, Oregon, treatment plant was initially designed in 1969 with a 1-minute flash mixing basin followed by a 75-minute contact basin. The plant flow sheet is as illustrated in Figure III-4. Filtration is provided by four tri-mixed media gravity filters rated at 5 gpm/ft^2. Raw water turbidity varies from 5 to 1,000 NTU but generally averages less than 10 NTU. Filtered water turbidity ranges from 0.1 to 0.2 NTU with filter runs reaching 30 hours in the winter to as much as 70 hours during the summer, during conditions of lower raw water turbidity.

Virginia--Direct filtration as a treatment method for water from five sources was evaluated with pilot test equipment in Virginia.[19] The most effective filtration scheme consisted of a 3-minute rapid mix with alum and a cationic polymer used in combination as primary coagulants. The rapid mix was followed by filtration at 5 gpm/ft^2 through 20 inches of 1.3 mm effective size anthracite coal and 10 inches of 0.45 mm effective size silica sand. This configuration produced filtered water of 0.1 NTU or less with filter run times of at least 8 hours when raw water was less than 15 to 20 units for both color and turbidity. A filtered water turbidity of 0.1 NTU resulted in practically complete removal of algae and coliform bacteria.

Erie County--The existing facilities at the Sturgeon Point filter plant of the Erie County Water Authority (New York) permitted the side-by-side comparison of direct filtration and conventional (pretreatment-filtration) treatment of raw water from Lake Erie with a minimum of modifications.[20] The goal of the study was to determine whether or not direct filtration would produce water of comparably high quality to that now being produced by the plant, without resulting in undesirable operating conditions. For evaluation of direct filtration, the raw water stream to the treatment units was split, permitting full-scale comparison of the two treatment processes on the same raw water.

The following conclusions were drawn from the results of the study program:

- The direct filtration process can consistently produce high quality filtered water. At filtration rates of 2 to 6 gpm/ft^2 direct filtration produces filtered water with a turbidity of 0.1 to 0.3 NTU, comparable to that produced by conventional treatment. During the study period, raw water turbidities varied from 1 to 100 NTU. Over the two-year study period the raw water turbidity averaged 8.6 NTU.

- Direct filtration reduced chemical costs by decreasing the required alum dosage from 15 to 5 mg/L.

Two Harbors-- Two Harbors, Minnesota, was one of the north shore Lake Superior communities that constructed water filtration plants after amphibole asbestos fibers were found in their drinking water in 1973. The 6 mgd Two Harbors plant, completed in 1977, provides direct filtration with flocculation.[21] Plant processes include two-stage rapid mix (8.5 minutes mix time at design flow, G = 200 sec^{-1}), two-stage flocculation (38.5 minutes flocculation time at G = 20 sec^{-1}), and gravity filtration at 4 gpm/ft^2 through tri-media filters consisting of anthracite (16.5 inches), sand (9 inches), and illmenite (4.5 inches).[22] Chemical addition, including chlorine, nonionic polymer, and 12 mg/L alum, is provided at the rapid mix basins.

The turbidity of Lake Superior has always been generally excellent. For samples taken between 1978 and 1981, raw water turbidity ranged between 0.3 NTU and 1.5 NTU (0.8 NTU average). During the same period, filtered water samples ranged from 0.03 NTU to 0.20 NTU (0.12 NTU average), with an average turbidity removal of 83 percent.[22] Substantial reductions in fiber count are consistently achieved.

Portola--The City of Portola, Califorina, treats water from a reservoir with a pressure-filtration plant.[23] The 1.6 mgd installation has two 8-foot diameter horizontal pressure filters containing tri-media rated at 5 gpm/ft^2. Raw water is prechlorinated, coagulated with 15 to 20 mg/L of alum, and filtered. Powdered activated carbon is often added ahead of the filters for taste and odor control. Lake water turbidity averages 2 to 3 NTU during the summer when this facility is operated. Filtered water turbidities range from 0.05 to 0.5 NTU, but generally average 0.1 to 0.2 NTU. Because of the relatively heavy turbidity loading and summer algae problems, filter runs range from 5 to 6 hours.

Diatomaceous Earth Filtration

Process Description--

Diatomaceous earth (DE) filtration, also known as precoat or diatomite filtration, is applicable to direct treatment of surface waters for removal of relatively low levels of turbidity. Diatomite filters were developed by the U.S. military forces during World War II to remove cysts of Entamoeba histolytica from water. After the war, diatomite filters were applied to civilian use, principally in the filtration of swimming pool water. A few were constructed to treat municipal supplies. Many of these installations were unsatisfactory because the basic principles of diatomite filtration were not understood. In many instances, the lack of adequate pretreatment led to failure of the system.

Diatomite filters consist of a layer of DE about 1/8-inch thick supported on a septum or filter element. The thin precoat layer of DE is subject to cracking and must be supplemented by a continuous-body feed of diatomite, which is used to maintain the porosity of the filter cake. If no body feed is added, the particles filtered out will build up on the surface of the filter cake and cause rapid increases in headloss. The problems inherent in maintaining a perfect film of DE between filtered and unfiltered water have restricted the use of diatomite filters for municipal purposes, except under certain favorable conditions. Figure III-5 shows a schematic of a typical pressure DE filtration system.

Currently there is interest in diatomite filters to remove Giardia cysts.[5] Research on G. lamblia cyst removal at the University of Washington showed that a DE filter could consistently remove more than 99 percent of the cysts.

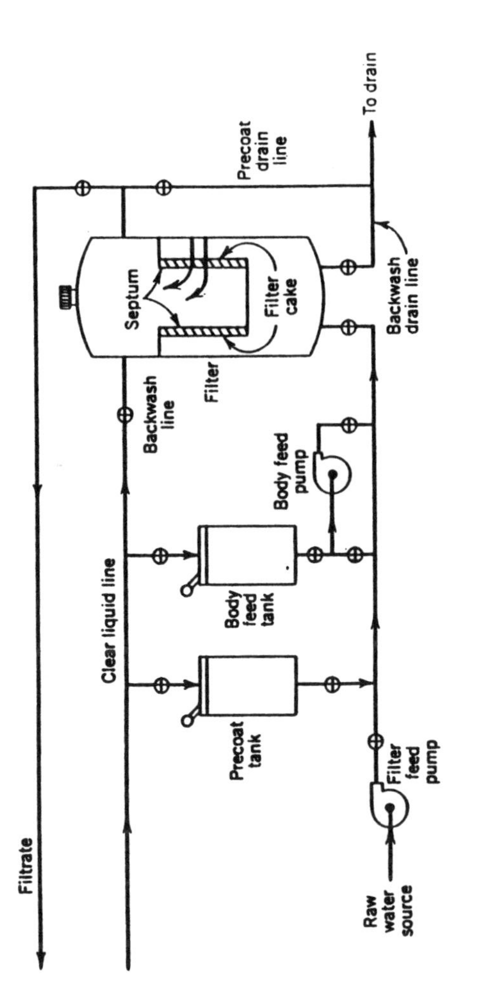

SOURCE: REFERENCE 24

Figure III-5. TYPICAL PRESSURE DIATOMACEOUS EARTH FILTRATION SYSTEM

Filtration with plain DE earth has shown an inability to remove very fine particles, including viruses.[9] Research has shown in some cases, however, that steps can be taken to achieve greater than 99 percent virus removal.[25]

The potential applications as well as the limitations of these filters is well described by E.R. Baumann in an AWWA Task Group Report as follows:[26]

> The diatomite filter is an acceptable tool in the waterworks industry, but is only a step in the water-treatment process, the details of which must be predicated on the characteristics of the raw water and what must be done to condition the water prior to filtration. Present municipal experience relates principally to turbidity removal from surface waters. Municipal experience with turbidity removal has been principally where the actual turbidity is relatively low and the bacteriological quality is good, thus little pretreatment is provided. There is no common agreement on the upper limits of turbidity that can be handled without pretreatment. There has been some extension of the use of diatomite filters in some of the areas (i.e., marginally, moderately, and grossly polluted surface waters) which have been of particular concern to regulatory public health personnel. No one should recommend diatomite filtration in the near future for filtration of a grossly or even moderately polluted supply. However, appropriate prior conditioning of such waters by one of several available means, including coagulation and settling or chemical treatment of the filter medium or water to improve water filterability, will undoubtedly lead to wider application of diatomite filters. Currently, effective water conditioning coupled with filtration through carefully engineered filters, an appropriate filter aid body feed system, and with adequate disinfection should provide a substantial margin of safety for marginally polluted waters and even sometimes for moderately polluted waters.

An advantage of a diatomite filtration plant for potable water is the lower first cost of such a plant. On waters containing low suspended solids, the diatomite filter installation cost should be somewhat lower than the cost of either a conventional rapid-sand filtration plant or a direct filtration plant. Diatomite filters will thus find application in potable water treatment under the following conditions:

- In cases where the diatomite plant will be found to produce acceptable water at a lower total cost than any practical alternative.

- In cases where financial capacity is tightly circumscribed, the lower first cost of a diatomite filter installation may be the major factor in the final choice of plant.

- For emergency or standby service at locations experiencing large seasonal variations in water demand, the lower first cost of the diatomite filter may prove to be more economical.

Laboratory and Pilot Plant Studies--

A study was conducted at Colorado State University to determine the effectiveness of DE filtration for removal of Giardia cysts.[7] In addition, removals of turbidity, total coliform bacteria, standard plate count bacteria, and particles were determined. Parameters evaluated included type of DE, hydraulic loading rates, influent concentrations of bacteria and Giardia cysts, headloss, run time, temperature, and the use of alum-coated DE.

Hydraulic loading rates imposed were 1, 2, and 4 gpm/ft^2. Seven grades of DE were used. Temperatures varied from 5° to 19°C; concentrations of Giardia cysts ranged from 50 to 5,000 cysts/L; and bacteria densities varied from 100 to 10,000/100 mL.

The results of this study showed that DE filtration is an effective process for water treatment. Giardia cyst removals were greater than 99.9 percent for all grades of DE tested, for hydraulic loading rates of 1.0 to 4.0 gpm/ft^2, and for all temperatures tested. Percent reduction in total coliform bacteria, standard plate count bacteria, and turbidity are influenced strongly by the grade of DE. The coarsest grades of DE recommended for water treatment will remove greater than 99.9 percent of Giardia cysts, 95 percent of cyst-sized particles, 20 to 35 percent of coliform bacteria, 40 to 70 percent of heterotrophic bacteria, and 12 to 16 percent of the turbidity from Horsetooth Reservoir water. The use of the finest grade of DE or alum coating on the coarse grades will increase the effectiveness of the process, resulting in 99.9 percent removals of bacteria and 98 percent removals of turbidity.

DE filtration was evaluated at McIndoe Falls, Vermont, during parallel studies with slow sand filtration.[27] Filtration rates averaged 1 to 1.8 gpm/ft^2 on the 10 to 20 gpm pilot pressure unit. The key conclusions from this study are as follows:

- Pressure DE filtration removed Giardia cysts dependably, providing 99.97 percent reduction.

- Total coliforms were reduced 86 percent or more in 70 percent of the samples, and standard plate count bacteria were reduced 80 percent or more in 70 percent of the samples.

- The average bacterial content in the effluent under ambient conditions was 38/100 mL for total coliforms and 6/mL for standard plate count bacteria.

Malina et al, reported that a high percentage of removal could be attained for poliovirus when coated DE filter aid was used or when cationic polymer was added to the raw water.[25] In one 12-hour filter run, DE earth coated with 1 mg of cationic polymer per gram of DE, produced filtered water in which no viruses were recovered from 11 samples (removal >99.95%). In another run, one of 12 samples was positive, and in this instance, virus removal was 99 percent. In a 12-hour run in which uncoated DE was used, but 0.14 mg/l of cationic polymer was added to the raw water, no viruses were recovered from any of the 12 samples analyzed.

Slow-Sand Filtration

Process Description--

Slow-sand filters are similar to single-media rapid-rate filters in some respects, yet they differ in a number of important characteristics. In addition to the obvious difference of flow rate, slow-sand filters: (1) function using biological mechanisms instead of physical-chemical mechanisms, (2) have smaller pores between sand particles, (3) do not require backwashing, (4) have longer run times between cleaning, and (5) require a ripening period at the beginning of each run.

Recently there has been an increased interest in slow-sand filters. This interest has been expressed principally in small communities which have a protected surface watershed with only chlorination. The need for multiple barriers has been demonstrated for protection against giardiasis. Slow-sand filters are attractive to small water systems since they require little operator attention and no chemical pretreatment. For proper application of slow-sand filters, the raw water must be of high quality (less than 10 NTU with no color problem).

Use of slow-sand filters can sometimes be prohibitive because of the large land areas required. Another disadvantage is the difficulty of achieving good results under all raw water conditions. Slow-sand filters are often covered to protect against freezing in winter and algae growths in summer. Very few slow-sand filters have been built in the United States since 1915.

The rate of filtration varies from 1 to 10 million gallons per acre per day (mgad) with 3 to 6 mgad (0.05 to 0.1 gpm/ft^2) the usual range. Filter sand depth ranges up to 42 inches. Cleaning is accomplished by scraping off 1 or 2 inches from the surface. Typically, once the depth is reduced to 24 inches new sand is added. The sand has an effective size of 0.25 to 0.35 mm, and a uniformity coefficient of 2 to 3. The underdrainage system usually is constructed from split tile, with laterals laid in coarse stone and discharging into a tile or concrete main drain. Slow-sand filters constructed recently include perforated PVC pipe for laterals. The initial loss of head is only about 0.2 feet. When the headloss reaches about 4 feet the surface is usually scraped. The length of run between cleanings is normally 20 to 60 days.[28] With varying combinations of raw water quality, sand size, and filtration rate, however, runs may be shorter or longer than are normally experienced.

Many slow-sand filters have no pretreatment, while others are preceded by coagulation, settling, or roughing filters. Slow-sand filters are cleaned by scraping a surface layer of sand and washing the removed sand, or washing the surface sand in place by a traveling washer. Since slow-sand filters must be removed from service for extended time periods for cleaning, redundant or standby filters are needed.

As noted above, slow-sand filters produce poorer quality filtrate at the beginning of a run (right after scraping), and require a filter-to-waste (or ripening) period of 1 to 2 days before being used to supply the system.[10] A ripening period is an interval of time immediately after a scraped or resanded filter is put back on-line, in which the turbidity or particle count results are significantly higher than the corresponding values for a control filter. More recent work indicates that scraping does not significantly affect Giardia removal, as long as the sand bed has developed a mature microbiological population.[18,8]

Laboratory and Pilot Plant Studies--

A 14-month pilot study evaluated the effectiveness of a slow-sand filter to remove turbidity and coliform bacteria from a surface water supply.[10] The pilot filter contained a 37-inch deep, 0.32 mm effective size sand bed and was operated at a filtration rate of 0.05 gpm/ft^2 (3 mgad). After an initial 2-day ripening period, effluent turbidity was consistently near or below 0.1 NTU at applied water turbidities which averaged 4 to 5 NTU. Coliform bacteria removal was always 99.4 percent or better reaching 100 percent in one filter run. Removal of Giardia cyst-sized particles (7 to 12 µm size range) averaged 96.8 percent or better for all test runs. Even though Giardia removal was not measured directly, the results of this study clearly established the suitability of slow-sand filtration as a viable technology for producing high quality filtered water.[10] Other studies, like those described below, have confirmed that the biological action in a sand bed adds significantly to Giardia removal, so that higher removal rates than those found for inorganic particles can be expected.

Treatment efficiency of slow-sand filtration was studied under various design and operating conditions to ascertain removal of Giardia lamblia cysts, total coliform bacteria, standard plate count bacteria, particles, and turbidity.[18,8] Filter removals were assessed at hydraulic loading rates of 1 mgad, 3 mgad, and 10 mgad; temperatures of 0°, 5°, and 17°C; effective sand sizes of 0.128, 0.278

and 0.615 mm; sand bed depths of 0.48 and 0.97 mm; influent Giardia cyst concentrations of 50 to 5,000 cysts/liter; and various conditions of filter biological maturity and influent bacteria concentrations.

Results showed that slow-sand filtration is effective in removing microbiological contaminants. Giardia cyst removal was consistently greater than 99.8 percent for a biologically mature fully ripened filter. Total and fecal coliform removal was approximately 99 percent. Particle removal averaged 98 percent. Standard plate count bacteria removal ranged from negative removals to 99 percent, depending on the influent concentration. Turbidity displayed a unique ability to pass through the filters, a characteristic not previously reported, and removal ranged from 0 to 40 percent. It is entirely possible that the particles measured as turbidity in this water supply were too small to be captured by the filter or perhaps were charged such that they were repelled by the filter media and passed through in the effluent. Some of the turbidity could also be due to debris sloughing off the filter. Operating results from Waverly, New York, tend to substantiate the former possibilities.[29]

Changes in process variables resulted in decreased coliform removal efficiency for increased hydraulic loading rate, increased sand size, decreased bed depth, and decreased biological activity. Giardia removal was influenced by the biological maturity of the filter, but not by the variables mentioned above. During filter start-up, Giardia removal was 98 percent; and once the filter was mature, removal was virtually complete.[18,8]

Case Histories--

Denver Water Board--The Denver Water Board (Colorado) operates a 10.5 acre slow-sand filter, built in 1901, to supplement the capacity of their other treatment plants.[30] This facility which processes water from the North Platte River, operates at flows ranging from 7 to 45 mgd.[30]

The six filter basins contain a 40-inch deep filter bed consisting of 0.25 to 0.55 mm effective size sand with a sand uniformity coefficient of 2.7 to 3.3. The sand is obtained from the riverbank and processed to the proper size on-site.

Platte River turbidities range from 30 to 100+ NTU and presedimentation is provided before the water is applied to the filters. The turbidity of the settled water generally does not exceed 10 NTU. The best performance is obtained at filtration rates of 1.2 to 3 mgad. Filtered water turbidities range from 0.1 to 0.2 NTU. At the lower rates, turbidities are often less than 0.1 NTU.

Filter-run lengths are variable. Under light loading conditions at flows of 1 to 1.5 mgad, operating cycles between filter cleanings approach 90 to 100 days. At the higher flows, the operating cycles may be as short as 10 to 15 days.

Cleaning is labor-intensive. The top 1/2 inch of the sand bed is removed by a specially constructed device, which deposits the sand in windrows. The windrows are picked up by a loader, conveyed into a sand washing hopper, cleaned and returned to the filter.

Survey of 27 Plants--A survey of 27 slow-sand filtration plants in the United States indicated that most of the plants serve fewer than 10,000 persons, are more than 50 years old, and are effective and economical to operate.[31] Most facilities surveyed used lakes or reservoirs as raw water sources. Filtration rates ranged from less than 0.3 mgad to about 13 mgad. Fifty percent of the plants have filtration rates in the 2.6 to 6.4 mgad range. Filter media depths fall between 15 and 72 inches. Most installations use sieved sand with effective sizes averaging 0.2 to 0.4 mm. Sand uniformity coefficients varied from 1.4 to 3.5.

Figure III-6 presents average raw water turbidities for the slow-sand filter plants surveyed. About 50 percent of the plants process raw water with an average turbidity exceeding 2 NTU. All plants surveyed treat raw water with turbidities of less than 10 to 12 NTU.

Figure III-7 shows that 85 percent of the plants produce average filtered water with less than 1 NTU, and 50 percent average less than 0.4 NTU. One plant reported average filtered effluent turbidities of less than 0.1 NTU. Average annual coliform concentrations in the raw water are shown in Figure III-8. Figure III-9 shows for the three plants reporting that 90 percent of the facilities maintained effluent coliform levels of 1/100 mL or less.

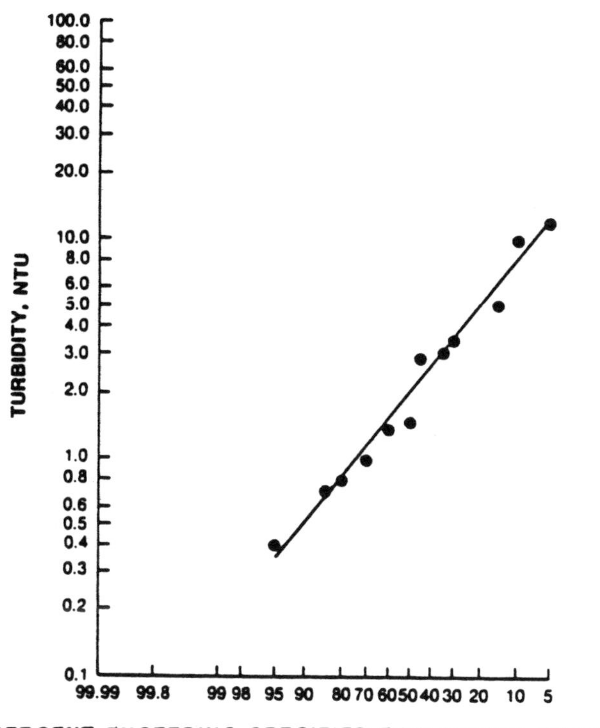

SOURCE: REFERENCE 31

Figure III-6. AVERAGE RAW WATER TURBIDITIES AT SLOW-SAND FILTER PLANTS SURVEYED

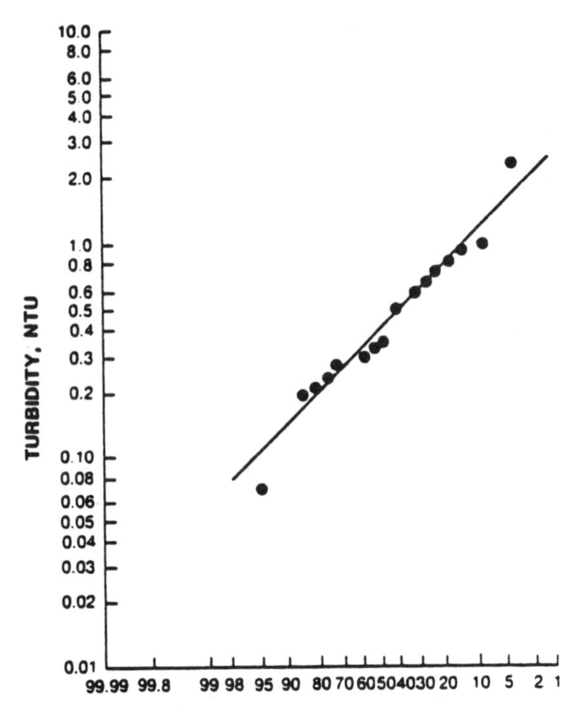

SOURCE: REFERENCE 31

Figure III-7. AVERAGE FILTERED WATER TURBIDITIES AT SLOW-SAND FILTER PLANTS SURVEYED

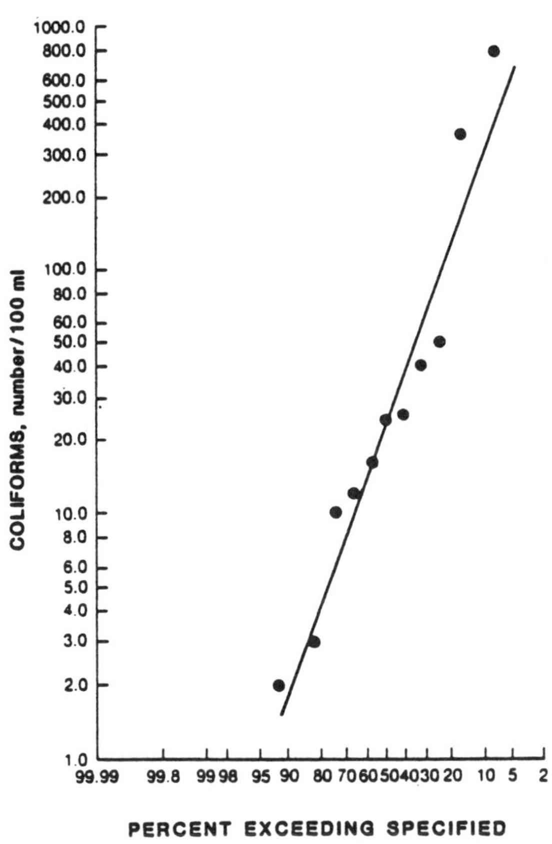

SOURCE: REFERENCE 31

Figure III-8. AVERAGE COLIFORMS IN RAW WATERS AT SLOW-SAND FILTER PLANTS SURVEYED

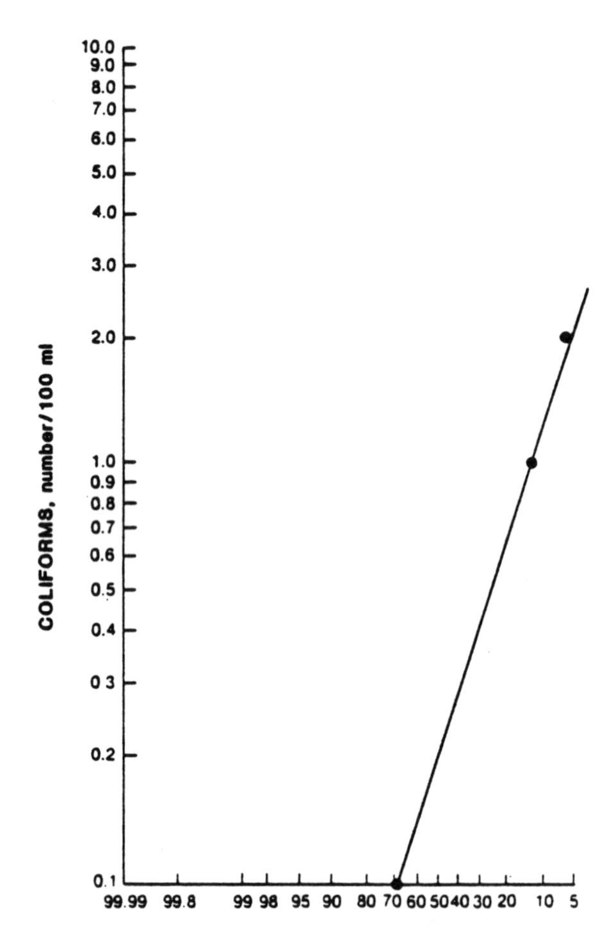

SOURCE: REFERENCE 31

Figure III-9. AVERAGE COLIFORMS IN FILTERED WATERS AT SLOW-SAND FILTER PLANTS SURVEYED

Filter cycle durations vary seasonally. The mean length of filter cycles ranged from 60 days in winter to 42 days in spring.

New York State--A study was performed at seven slow-sand filter installations in New York to assess the impact on filtered water quality of the cleaning procedure which involves scraping a thin layer of sand from the surface of the filter.[29] The performance of each filter was monitored before and after the scraping procedure. Effluent samples were analyzed for turbidity, total particle count, standard plate count, and total coliforms. Table III-5 lists the characteristic features of the seven facilities.

TABLE III-5. CHARACTERISTICS OF SLOW-SAND FILTER INSTALLATIONS IN NEW YORK[29]

Location	Slow-Sand Filtration Average Operating Flow rate (mgd)	Design Filtration Rate (mgad)	Average Operating Filtration Rate (mgad)	Filter Sand	
				Uniformity Coefficient	Effective Size (mm)
Auburn	6.0	2.83	3.6	2.4	0.45
Geneva	2.5	4.9	4.9	1.9	0.37
Hamilton	~0.3	---	1.0	2.4	0.27
Ilton	1.5	---	4.1-4.6	2.2	0.37
Newark	2.0	4.1	4.1	1.7	0.35
Ogdensburg	3.6	5.1	4.6	1.7	0.35
Waverly	1.2	4.1	4.1	2.4	0.15

Table III-6 is a summary of the data collected during plant tests designed to determine the impact of filter scraping on filtered water quality during the ripening period.

It is noted in Table III-6, that even during the ripening period, filtered effluent turbidities at all but the Waverly installation did not exceed 0.43 NTU. At the Waverly facility, it appears that the raw water contains submicron particles

TABLE III-6. FILTER RIPENING DATA - SUMMARY

Location	Type of Operation During Visit*	Date of Site Visit	Raw Water Turbidity During Site Visit (NTU)	Water Temperature During Site Visit	Filter Turbidity Approximately 5 Hours After Filter Startup (NTU)		Evidence of Ripening Period	Approximate Length of Ripening Period (days)
					Scraped/ Resanded Filter	Control** Filter		
Auburn	(1)	July 83	1.2-2.0	~19°C	0.43	0.27	Yes	0.25
Auburn	(1)	July 83	1.2-2.0	~19°C	0.28	0.27	None	--
Auburn	(1)	July 84	2.0-2.8	~18°C	0.22	0.23	None	--
Geneva	(1)	July 83	-	-	-	-	-	--
Hamilton	(1)	May 84	1.0-1.5	~12°C	0.28	None	None	--
Ilion	(1)	July 83	2.0-4.0	~23°C	0.30	0.40	minimal (particle count only)	0.5
Newark	(1)	Aug. 83	1.2-3.5	~13°C	0.35	0.35	None	--
Newark	(2)	Jan. 84	0.6-2.7	~ 4°C	0.41	0.12	Yes	2
Ogdensburg	(3)	Aug. 83	0.3-0.6	~15°C	0.12	0.10	None	--
Ogdensburg	(3)	Feb. 84	1.0-1.2	~ 2°C	0.22	0.24	None	--
Waverly	(1)	June 84	6.0-11.0	~15°C	2.3	1.6	Yes	10

*(1) Scraping operation.
(2) Resanding operation.
(3) Scraping combined with resanding.

**Control filter - Filter on-line at least one month, except Ogdensburg where the filter was on-line one week.

which scatter light and increase the turbidity, but are not efficiently removed by slow-sand filtration. According to the particle count data, Waverly removes particles larger than 2 µm as efficiently as the other plants visited.

Turbidity data from the control filters which have been in service for at least one month, typifies the performance of these facilities.

The study included a comprehensive evaluation of total particle count, standard plate count, and coliform removal from both the ripening filter and a control filter. In general, removal of total particles in the size range 2 µm to 60 µm measured by a total particle counter, ranged from 90 to 99.8 percent. Filtration generally reduced the total plate count, however, results were widely variable. Removals ranged from greater than 50 percent to no removals. Coliform removals through the ripening and control filter were only measured at two facilities. One facility achieved 98.5 percent removal of coliform organisms after completion of the ripening period.

McIndoe Falls--A two-year study at McIndoe Falls, Vermont, was conducted to evaluate the effectiveness of a slow-sand filter for removal of turbidity, bacteria, coliforms, and Giardia cysts.[27] The community obtains its water from two impoundments which capture water from two springs. The ponds are shallow and contain several beaver dams and lodges. Because of the presence of beavers, giardiasis was a major concern. Raw water quality is generally quite good with turbidities ranging from 0.4 to 4.6 NTU with a seasonal average of 2.1 NTU.

The two filters have an individual area of 400 square feet. They contain 42 inches of 0.33 mm effective size silica sand and have a design filtration rate of 2.05 mgad. During the evaluation, the following results were achieved:

- Removed turbidity to less than 1 NTU, 99.19 percent of the time (after the first 100 days of operation, the effluent values were below 1 NTU, 99.68 percent of the time) and 72 percent of the time the values were 0.2 NTU or less (raw water 1.45 NTU or less, 72 percent of the time).

- Reduced total coliform to 10/100 mL or less, 86 percent of the time (raw water 1300/100 mL or less, 86 percent of the time) and standard plate count to 10/mL or less, 94 percent of the time (raw water 500/mL or less, 94 percent of the time) under ambient load conditions.

- The average microbial content in the filter effluent under ambient conditions was 4 organisms/100 mL (raw water 440/100 mL) for total coliform, and 15 heterotrophic organisms per mL (raw water 520/mL) for standard plate count.

- Removed massive spikes of total coliform and standard plate count bacteria from raw water at temperature conditions above the range of 5 to 10°C.

- Did not remove bacteria as efficiently at temperatures below 5°C and particularly around 0 to 1°C. Spiking studies demonstrated the temperature effect with removal of total coliform deteriorating from 98 to 43 percent, and standard plate count from 98 to 80 percent in 9 days at 1°C. At 6 to 11°C, removal remained at an average of 99 percent for total coliform and 97 percent for standard plate count during a 15-day spike.

- Removed Giardia cysts very dependably, 99.98 percent or better, under warm temperature conditions.

- Did not remove Giardia cysts as completely at low temperatures. Under 7°C, 99.36 to 99.91 percent removal was achieved.

- During cold water conditions (below 5°C and particularly around 0° to 1°C) the biologic treatment process in the slow-sand filter was less effective in removing bacteria and Giardia cysts. Cyst removal was reduced to 93.7 percent, and total coliform and standard plate count removals dropped to 43 percent and 79 to 82 percent, respectively.

Package Plants

Process Description--

Package plants are not a separate technology in principle from the preceding technologies. They are, however, different enough in design criteria, operation, and maintenance requirements that they are discussed separately in this document.

Package plants can be used to treat water supplies for communities as well as recreational areas, state parks, construction camps, ski resorts, remote military installations and others where potable water is not available from a municipal supply. Several state agencies have mounted package plants on trailers for emergency water treatment. Their compact size, low cost, minimal installation requirements and ability to operate with less attention than larger, custom-designed facilities, makes them an attractive option in locations where revenues are not sufficient to pay for a full-time operator. Operator requirements vary significantly, of course, with specific situations, and under unfavorable raw water conditions they could demand full-time attention.

The package plant is designed as a factory-assembled, skid-mounted unit generally incorporating a single, or at the most, several tanks. A complete treatment process typically consists of chemical coagulation, flocculation, settling and filtration. Package plants, for purposes of this document, generally can be applied to flows ranging from about 25,000 gpd to approximately 6 mgd.

Package plants are most widely used to treat surface supplies for removal of turbidity, color and coliform organisms prior to disinfection. A limited application for package plants is treatment of well water for removal of iron, manganese, and hardness. Additionally, package plants can also remove many inorganic chemicals for which MCL's have been established. Inorganic contaminants such as arsenic, cadmium, chromium, lead, inorganic and organic mercury, selenium, silver and fluoride can be partially or totally removed under proper treatment conditions by chemical coagulation followed by filtration.

Package plants are also effective in removing contaminants which relate to the aesthetics of drinking water. Although organic color may be treated adequately by simple chlorination, it can be coagulated with alum or iron salts and removed in package plants by filtration, producing a water of more appealing quality. Organics contributing to taste and odor in drinking water supplies can also be removed by adsorption on powdered activated carbon followed by filtration to remove the residual carbon. In this regard, package plants have application to treatment of a wide range of water supplies and can improve greatly the safety and the overall acceptance of the finished water to the consumer.

Package Plant Performance--

The widespread acceptance of package plants as an economical solution to water treatment needs of small systems has resulted in construction of a significant number of plants during the 1970's. The quality of water produced by these plants was of concern and resulted in an on-site investigation at six selected facilities.[32] The six selected plants were in year-round operation, used surface water sources, and served small populations. Plants were monitored to assess the performance and ability to supply water meeting the interim primary drinking water regulations.

Only two of the plants consistently met the 1 NTU effluent standard. Effluent turbidities from one of the two plants were also less than 0.5 NTU in eight out of nine samples, while the second plant produced water lower than 0.5 NTU on six out of ten samples. These plants obtained their raw water from a relatively high quality source. A third plant met the turbidity standard on six out of nine visits, with values less than 0.5 NTU on five of the six visits. The other three plants met the turbidity standard on fewer than one-half the visits. Table III-7 compares raw to filtered effluent turbidities measured during on-site visits.

According to the authors of the survey, the performance difficulties of plants P, V, and M were related to the short detention time inherent in the design of the treatment units, the lack of skilled operators with sufficient time to devote to operating the treatment facilities and (in the cases of plants V and M), the wide-ranging variability and quality of the raw water source. It is reported that the raw water turbidity at the site of plant V often exceeded 100 NTU. Later, improvement in operational techniques and methods resulted in substantial improvement in effluent quality. After the adjustments were made, plant filters were capable of producing a filtered water with turbidities less than 1 NTU even when influent turbidities increased from 17 to 100 NTU within a 2-hour period.

One of the major conclusions of this survey was that package water treatment plants manned by competent operators can consistently remove turbidity and bacteria from surface waters of a fairly uniform quality. Package plants applied where raw water turbidities are variable require a high degree of operational skill and nearly constant attention by the operators. Further, it was pointed out that

TABLE III-7. PLANT TURBIDITY VALUES (NTU)*

Plant C		Plant W		Plant T		Plant V		Plant M		Plant P	
Raw	Clearwell Effluent	Raw	Clearwell Effluent	Raw	Clearwell Effluent	Raw	Clearwell Effluent	Raw	Clearwell Effluent	Raw	Clearwell Effluent
8.5	0.3	--	0.9	10.0	1.9	4.0	1.8	--	0.2	12.0	0.8
6.2	0.2	5.0	0.3	8.0	0.2	12.0	2.8	39.0	3.8	4.4	2.4
1.2	0.3	4.2	0.4	6.0	0.4	-- †	9.6	40.0	2.6	--	7.0
1.6	0.1	19.0	0.8	3.2	1.1	35.0	1.5	27.0	2.4	3.5	1.5
2.2	0.1	9.2	2.0	3.2	0.2	42.0†	2.0	6.0	1.2	2.0	0.1
4.0	0.1	11.5	0.3	3.2	0.2	10.0†	2.4	3.8	0.1	1.2	0.5
12.6	0.7	12.0	0.2	5.8	0.2	90.0†	8.5	73.0	11.0	15.6	9.7
5.2	0.2	11.0	0.3	10.4	3.2	28.0†	5.4	3.6†	0.1	3.1	2.2
2.2	0.2	29.7	0.9	3.4	0.7	19.0†	0.3	3.8	0.3	17.2	1.9
		12.8	0.2			47.0†	1.2	6.0	0.5		
						13.0†	0.8	70.0†	16.0		
						8.0†	0.3	25.0†	3.4		
						6.0†	0.3	>100.0†	55.0		
						>100.0†	0.5	>100.0†	31.0		
						60.0†	0.5	8.5†	2.2		
						24.0	1.2	4.3	0.4		
						13.0	0.3	4.0†	1.0		
						2.7	1.2	9.6†	1.9		
						1.2	1.0	19.1	1.1		
						3.3	0.5	64.0	6.9		
								8.2	1.0		

*Reference 32
†Averaged Values for Day

regardless of the quality of the raw water source, all package plants require a minimum level of maintenance and operational skill if they are to produce satisfactory water quality.[32]

The operation and performance of 36 package water treatment plants were surveyed in a study completed in 1979.[33] The capacity of these plants ranged from 28,800 gpd to 1.4 mgd and they served either municipalities or recreational sites. Raw water was predominantly from surface sources; however, several plants treated groundwater for iron and manganese removal. Fourteen of the 36 plants treated raw water with turbidities less than 5 NTU. Two plants processed surface water where the turbidity exceeded 100 NTU.

Turbidity standards were being met at 23 of the 31 plants taking turbidity data during the survey. Eight did not meet the existing 1 NTU standard. Coliforms were detected in filtered water at 3 of the 31 plants in operation. Only one plant had coliforms in significant concentrations and had no measurable chlorine residual in the treated water. The two plants exceeding a raw water turbidity of 100 NTU had finished waters less than 1 NTU. Based upon the information collected in this survey, it was concluded that package plants are capable, when operating properly, of consistently providing finished water with turbidities less than 0.5 NTU.

A more recent survey of 27 package water treatment plants was made in 1986 to evaluate the performance of the plants in removing turbidity.[34] Results of that survey showed that all 27 plants had average turbidity values less than 1.0 NTU, and 18 of the plants produced filtered water with an average turbidity of 0.5 NTU or less.

Colorado State University conducted a series of tests on one package plant over a 5-month period during the winter of 1985-86.[35] Existing installations in Colorado had proven effective for turbidity removal, and the tests at the university were designed to evaluate the systems effectiveness in removing coliform bacteria and _Giardia_ cysts from two low turbidity, low temperature source waters. The test results showed that the filtration system could remove greater than 99 percent of _Giardia_ cysts for waters having less than 1 NTU turbidity and less than 5°C temperatures, so long as proper chemical treatment is applied and the unit is operated at 10 gpm/ft^2 or less. In addition, for Horsetooth Reservoir source water, effluent turbidities were consistently less than 0.5 NTU during 12 separate runs when coagulant doses were between 15 mg/L and 45 mg/L. The source water turbidity varied between 3.9 and 4.5 NTU for these test runs.

IV. Disinfection in Community Systems

GENERAL

While the filtration processes described in Section III are intended to physically remove microbial contaminants from water supplies, disinfection is specifically used to inactivate or kill these organisms. Sterilization, or the complete destruction of all organisms in water, is not considered within the scope of this discussion. Disinfection is most commonly achieved by adding oxidizing chemicals to water, but can also be accomplished by physical methods (applying heat or light), by adding metal ions, or by exposure to radioactivity.

Disinfection is always the final process in a water treatment system, and often is the only treatment given to some supplies. Predisinfection is practiced in some plants to control growth of algae and microorganisms in the plant. The efficiency of disinfection processes depends directly on the clarity of the water being treated. In most cases, therefore, effective disinfection of surface waters requires effective filtration as preliminary treatment, as described in Section III. Disinfection of turbid waters is difficult, inefficient, and generally impractical. Minimum water turbidities (less than 0.5 NTU) are necessary to assure maximum contact between pathogens in the water and the disinfectant added.

Many disinfectants and disinfection processes have been used in treatment of water supplies, and new ones are continuously being proposed for use. Not all of these have been effective and practical. Fair et al have recommended the following criteria for evaluating any potential disinfectant:[1]

- Ability of the disinfectant to destroy the kinds and numbers of organisms present within the contact time available, the range of water temperatures encountered, and the anticipated fluctuations in composition, concentration, and condition of the water being treated.

- Ready and dependable availability of the disinfectant at reasonable cost and in a form which can be conveniently, safely, and accurately applied.

- Ability of the disinfectant, in concentrations employed, to accomplish the desired objectives without rendering the water toxic or objectionable, aesthetically or otherwise, for the purposes it is intended.

- Ability of practical, duplicable, quick, and accurate assay techniques for determining disinfection concentration, for operating control of the treatment process, and as an indirect measure of disinfecting efficiency.

Three groups of technologies have been identified using the above criteria as general guidelines. Specific disinfection processes within the technology groups are listed in Section I and are discussed below in terms of their equipment requirements, design and operating parameters, performance, and applicability.

MOST APPLICABLE TECHNOLOGIES

The following methods of disinfection are identified as the Most Applicable Technologies (not necessarily in order of effectiveness), and are those most widely used for inactivation of microbial contaminants:

- Chlorination (chlorine liquid, gas, and hypochlorite)
- Chlorine dioxide
- Chloramination
- Ozonation

The performance of these and other chemical disinfectants can best be described through the use of the C·T product (the product of residual disinfectant, C, in mg/L, and contact time, T, in minutes). A detailed description of the application of the C·T concept to disinfection practice has been presented by Hoff.[2] The concept, based on Watson's Law ($k = C \cdot T$), has been used since 1962, although the background of the concept has not been widely explained. Recommendations of C·T values for disinfection practice make the implicit assumption that $n = 1$.

The range of concentrations and contact times for different disinfectants to achieve 99 percent inactivation of E. coli, poliovirus, and Giardia cysts are presented in Tables IV-1 and IV-2. As shown by the concentration-time products (C•T) in the tables, there is wide variation both in resistance of a specific organism to the different disinfectants, and in the disinfection requirements for different organisms using a single disinfectant. In general, however, the C•T products in the tables show that Giardia cysts are the most resistant to disinfection, followed by viruses, whereas E. coli are the least resistant.

TABLE IV-1. SUMMARY OF C•T VALUE RANGES FOR 99 PERCENT INACTIVATION OF VARIOUS MICROORGANISMS BY DISINFECTANTS AT 5°C

	Disinfectant			
Microorganism	Free Chlorine pH 6 to 7	Preformed Chloramine pH 8 to 9	Chlorine Dioxide pH 6 to 7	Ozone pH 6 to 7
E. coli	0.034-0.05	95-180	0.4-0.75	0.02
Polio 1	1.1-2.5	768-3,740	0.2-6.7	0.1-0.2
Rotavirus	0.01-0.05	3,806-6,476	0.2-2.1	0.006-0.06
Phage f_2	0.08-0.18	-	-	-
G. lamblia cysts	47->150	-	-	0.5-0.6
G. muris cysts	30-630	-	7.2-18.5	1.8-2.0

Source: Reference 2

TABLE IV-2. CONCENTRATION-CONTACT TIME OF DISINFECTANTS FOR 99 PERCENT INACTIVATION OF MICROORGANISMS

Microorganisms	Disinfectant	Concentration C, mg/L	Contact T, min	C·T	pH	Temperature °C
E. Coli[1]	Ozone (O_3)	0.065	0.33	0.022	7.2	1
		0.0023	1.03	0.002	7.0	12
	Chlorine Dioxide (ClO_2)	0.75	0.50	0.38	6.5	5
		0.75	0.30	0.23	6.5	10
	Hypochlorous Acid (HOCl)	0.1	0.4	0.04	6.0	5
	Hypochlorite ion (OCl^-)	1.0	0.92	0.92	10.0	5
	Dichloramine ($NHCl_2$)	1.0	5.5	5.5	4.5	15
	Monochloramine (NH_2Cl)	1.0	175	175	9.0	5
Poliovirus[1]	Ozone (O_3)	0.3	0.13	0.04	7.2	5
		0.245	0.50	0.12	7.0	24
	Chlorine Dioxide (ClO_2)	0.8	6.8	5.4	7.0	5
		0.5	2.0	1.0	7.0	25
	Hypochlorous Acid (HOCl)	0.5	2.1	1.05	6.0	5
	Hypochlorite ion (OCl^-)	0.5	21	10.5	10.0	5
	Dichloramine ($NHCl_2$)	100	140	14,000	4.5	5
	Monochloramine (NH_2Cl)	10	90	900	9.0	15
Giardia lamblia[2]	Free Chlorine	2.5	30	75	6	5
		2.5	47	118	7	5
		2.5	57	142	8	5

TABLE IV-2 (Continued)

Microorganisms	Disinfectant	Concentration C, mg/L	Contact T, min	C·T	pH	Temperature °C
Giardia lamblia[3]	Ozone (O_3)	0.15	0.97	0.15	7	25
		0.082	1.9	0.16	7	25
		0.034	5.5	0.19	7	25
		0.48	0.95	0.46	7	5
		0.20	3.2	0.64	7	5
		0.11	5.0	0.55	7	5
Giardia muris[3]	Ozone (O_3)	0.18	1.3	0.24	7	25
		0.10	2.2	0.22	7	25
		0.08	3.4	0.27	7	25
		0.70	2.5	1.8	7	5
		0.40	5.0	2.0	7	5
		0.31	6.4	2.0	7	5
Giardia lamblia[4]	Ozone (O_3)	0.03- 0.15	5.5-1.06	0.17	7	25
		0.11- 0.48	5.0-0.94	0.53	7	5
Giardia muris[5]	Chloramine	1.5 - 2.4	236-276	496	7	3
		1.4 - 2.9	122-227	354	7	10
		1.0 - 1.9	75-241	184	7	18
Giardia muris[6]	Chloramine	5.0 -16.6	50-182	848	7	15
		3.2 - 9.0	58-132	466	8	15

[1] Reference 3
[2] Reference 4
[3] Reference 5
[4] Reference 6
[5] Reference 7. Chloramines not preformed.
[6] Reference 78. Preformed chloramines.

The use of C•T values to interpret disinfection data has become more prevalent in the 1980's. The 99 percent inactivation level has been used for calculating C•T values in most studies, probably because it is the level at which exponential kinetics ($N/N = K \cdot T$) are usually best approximated. If exponential kinetics were followed, and if C•T values for 99 percent inactivation were known, C•T values for other levels of inactivation could easily be calculated. The ideal is not often observed, though, and great care must be used in any attempts at extrapolation.

The following paragraphs describe the methods and performance of each of the Most Applicable Technologies individually.

Chlorination

General--

The practice of chlorination has been used to control the outbreak of disease since its first continuous application to a New Jersey municipal water supply in 1908. For purposes of disinfection of municipal supplies, chlorine is applied primarily in two forms: as a gaseous element, or as a solid or liquid chlorine-containing hypochlorite compound. Gaseous chlorine is generally considered the least costly form of chlorine that can be used in large facilities. Chlorine is shipped in cylinders, tank cars, tank trucks, and barges as a liquified gas under pressure. Chlorine confined in a container may exist as a gas, as a liquid, or as a mixture of both. Thus, any consideration of liquid chlorine includes consideration of gaseous chlorine.

Hypochlorite forms (principally calcium or sodium) have been used primarily in small systems (less than 5,000 persons) or in large systems where safety concerns related to handling the gaseous form outweigh economic concerns. Present day commercial, high-test calcium hypochlorite products contain at least 70 percent available chlorine and are usually shipped in tablet or granular forms. Sodium hypochlorite is provided in solution form, containing 12 percent or 15 percent available chlorine.

When chlorine (Cl_2) is dissolved in water, it reacts to form hypochlorous and hydrochloric acids:

$$Cl_2 + H_2O \rightleftarrows HOCl + H^+ + Cl^-$$

This reaction is essentially complete within a few seconds. The hypochlorous acid ionizes or dissociates practically instantaneously into hydrogen and hypochlorite ions:

$$HOCl \rightleftarrows H^+ + OCl^-$$

These reactions represent the basis for use of chlorine in most sanitary applications.

HOCl and OCl- have considerably different capabilities for destruction of microorganisms, and therefore, it is important to know that the two forms exist in equal percentages (50-50) at about pH 7.5, that the percentage of HOCl increases nonlinearly (and OCl- decreases) as pH decreases, and that the reverse is true as pH increases.

Hypochlorite chlorine forms also ionize in water and yield hypochlorite ions which establishes equilibrium with hydrogen ions:

$$Ca(OCl)_2 + 2H_2O \rightleftarrows 2HOCl + Ca(OH)_2$$

$$NaOCl + H_2O \rightleftarrows HOCl + NaOH$$

Only recently has a negative aspect been realized regarding chlorination: under some conditions, chlorine reacts with certain organic substances present in some water supply sources to produce trihalomethanes (THM) and other by-products, which may have carcinogenic effects following continuous exposure over long periods of time. Exposure to these by-products has already been reduced to a great extent by changes in chlorination practice. These changes include the use of different points of application, the use of lower dosages, the use of chlorine in combination with other oxidants, and removal of precursor organics. In treating waters with high THM production potential where precursor removal is difficult or expensive, and where other chlorine disinfection by-products become of concern, the use of alternate disinfectants may replace or supplement chlorine in many instances.

Performance--
There are wide differences in the susceptibility of various pathogens to chlorine. The general order, from most susceptible to least susceptible, is (1) bacteria, (2) virus, (3) cysts, and (4) bacterial endospores.

The destruction of pathogens by chlorination is dependent on a number of factors, including water temperature, pH, contact time, degree of mixing, turbidity, presence of interfering substances, and concentration of chlorine available. Both pH and temperature have a marked effect on the rate of virus kill by chlorine. For example, several studies show that decreasing the pH from 7.0 to 6.0 reduced the required virus inactivation time by about 50 percent and that a rise in pH from 7.0 to 8.8 or 9.0 increased the inactivation period about six times.[9] An AWWA Committee reporting on virus in water concluded that "in the prechlorination of raw water, any enteric virus so far studied would be destroyed by a free chlorine residual of about 1.0 ppm, provided this concentration could be maintained for about 30 minutes and that the virus was not embedded in particulate material."[72] In a later AWWA committee report it was recommended that these conditions be maintained at a pH of less than 8 to ensure adequate protection from viruses.[76]

In general, disinfection by chlorination has been shown to be most efficient with relatively high values of chlorine residual, contact time, C•T products, water temperature, and degree of mixing; combined with relatively low values for pH, turbidity, and the presence of interfering substances. As indicated by the C•T product, it is possible to have excellent disinfection with low chlorine residuals, as long as long contact times are used and other factors are beneficial for disinfection. The converse is also true, i.e., high chlorine residuals with shorter contact times may also result in excellent disinfection, within practical ranges of residuals and contact times.

The following paragraphs present recent data regarding disinfection by chlorine from laboratory, pilot-plant, and full-scale water treatment plant sources. It is important to keep in mind the many differences between these three test environments when reviewing the information provided. Contact time, for example, is a critical factor in chlorination efficiency, and while laboratory conditions can sometimes approach theoretical contact times for plug-flow and complete-mixing conditions, plant-scale conditions in contact basins rarely do. Significant short-circuiting often occurs, reducing contact time, sometimes by as much as 95 percent.[10] Where uncertainty exists regarding the effect of short-circuiting, tracer studies would be needed to establish real disinfectant contact times.

Laboratory Research and Pilot-Plant Studies--Several relationships regarding susceptibility of pathogens to disinfection by chlorine, as well as the effects of varying pH are shown by laboratory data plotted in Figure IV-1. The bacteria E. coli and Shigella dysenteriae can be seen to require the lowest contact times and chlorine concentrations for destruction, followed by three viruses (polio, coxsackie A2, and hepatitis A). The Entamoeba histolytica cyst requires longer contact times and higher concentrations than the bacteria or viruses. The differences in susceptibility between species of the same type of organism are highlighted by the fact that the bacterium Bacillus anthracis (which forms a spore) is shown to require greater contact times and concentrations than the virus and cysts in this figure. The increased effectiveness of disinfection at lower pH values can be seen by comparing contact time-concentration curves for E. coli, coxsackievirus A2, E. histolytica, and B. anthracis.

The resistance of 20 different enteric viruses to free chlorine was studied by Liu et al.[12] These tests were all conducted under constant conditions of 0.5 mg/L free chlorine, a pH of 7.8, 2°C, using treated Potomac estuary water. As shown in Table IV-3, the least resistant virus was a reovirus requiring 2.7 minutes for a 99.99 percent (4 log) devitalization, and the most resistant was a poliovirus requiring more than 60 minutes for the same level of devitalization. Corresponding C•T factors for these 20 viruses range from 1.4 to more than 30 under the constant conditions of this work.

Virus survival tests have also been reported by Payment on a variety of both laboratory strains and field strains.[77] These tests were all conducted at a free chlorine residual of 0.4 mg/l, a pH of 7.0, and 5°C. Survival was analyzed at 10, 100, and 1,000 minutes of contact time. Test results (see Table IV-4) show that only two poliovirus strains of the total of 20 test cultures had reached a 99.99 percent inactivation after 10 minutes (C•T = 4), six poliovirus strains had reached 99.99 percent inactivation after 100 minutes (C•T = 40), and 11 of the 12 polio viruses plus one Coxsackievirus strain (12 out of 20 strains) had reached 99.99 percent inactivation after 1,000 minutes (C•T = 400).

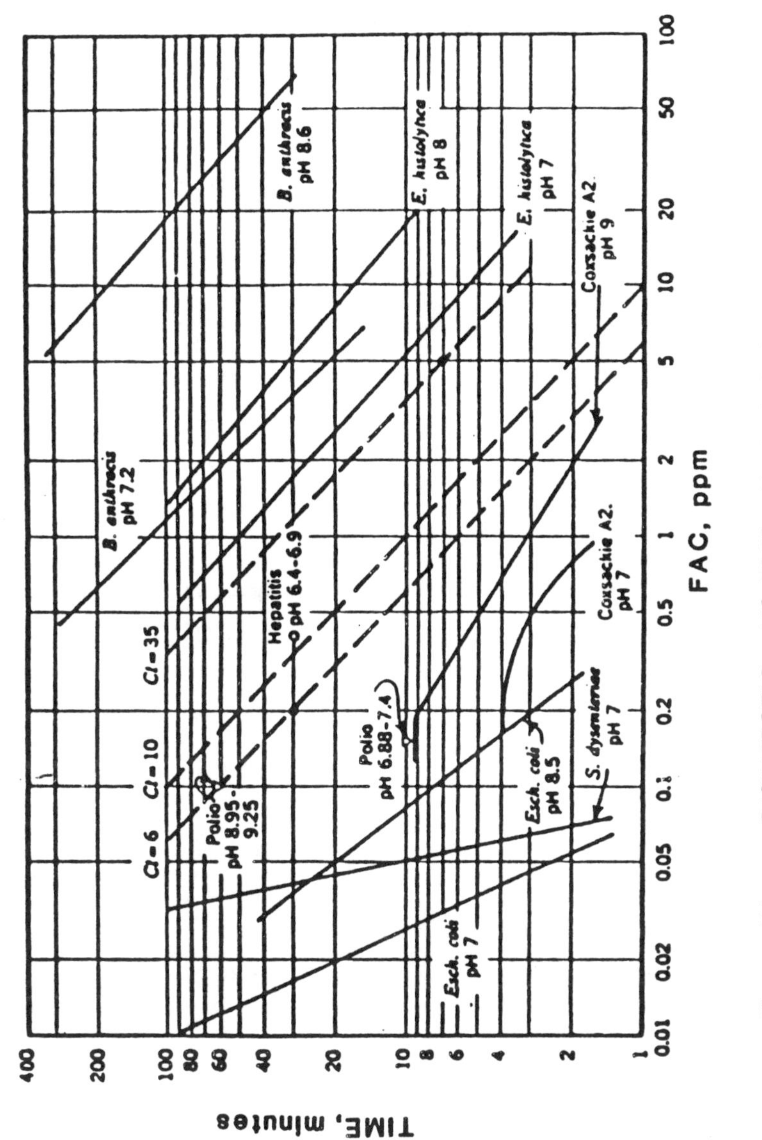

Figure IV-1. DISINFECTION VS FREE AVAILABLE CHLORINE RESIDUALS. TIME SCALE IS FOR 99.6-100% KILL. TEMPERATURE WAS IN THE RANGE 20-29°C, WITH pH AS INDICATED. (COURTESY AWWA[11])

TABLE IV-3. RELATIVE RESISTANCE OF TWENTY HUMAN ENTERIC VIRUSES TO 0.5 MG/L FREE CHLORINE IN POTOMAC WATER (pH 7.8 AND 2°C)

Comparison Based On				
	Percent Survival After 5 Minutes		Experimental Time for 99.99 Percent Deactivation	
	Virus	Percent	Virus	Time (min)
1.	Reo 3	<0.0009	Reo 1	2.7
2.	Reo 2	0.002	Reo 3	<4.0
3.	Reo 1	<0.005	Reo 2	4.2
4.	Adeno 3	<0.008	Adeno 3	<4.3
5.	Echo 7	0.135	Cox A9	6.8
6.	Cox A9	0.161	Echo 7	7.1
7.	Adeno 7a	0.330	Cox B1	8.5
8.	Adeno 12	0.330	Echo 9	12.4
9.	Polio 1	1.118	Adeno 7a	12.5
10.	Echo 29	1.660	Echo 11	13.4
11.	Echo 9	1.887	Polio 1	16.2
12.	Polio 3	2.420	Echo 29	20.0
13.	Cox B3	7.297	Adeno 12	23.5
14.	Cox B1	12.515	Echo 1	26.1
15.	Cox B5	14.533	Polio 3	30.0
16.	Echo 12	17.273	Cox B3	35.0
17.	Cox A5	18.620	Cox B5	39.5
18.	Polio 2	20.000	Polio 2	40.0
19.	Echo 11	20.833	Cox A5	53.5
20.	Echo 1	54.167	Echo 12	>60.0

Source: Reference 12

TABLE IV-4. VIRUS SURVIVAL AFTER 10, 100, AND 1,000 MINUTES OF CONTACT WITH AN INITIAL CONCENTRATION OF 0.4 MG OF FREE RESIDUAL CHLORINE PER LITER

Virus	% Survival after min of contact		
	10	100	1,000
Coxsackievirus B5 (no. 23, raw sewage)	70.00	21.67	0.079
Coxsackievirus B5 (no. 273, chlorinated water)	60.78	11.77	0.053
Coxsackievirus B5 (no. 241, chlorinated water)	3.43	0.24	0.041
Coxsackievirus B5 (laboratory strain)	1.44	1.22	<0.001
Coxsackievirus B4 (no. 1, treated sewage)	1.62	0.74	0.012
Coxsackievirus B4 (no. 358, chlorinated water)	0.31	0.063	0.014
Coxsackievirus B4 (no. 428, chlorinated water)	0.52	0.052	0.013
Coxsackievirus B4 (laboratory strain)	0.79	0.025	0.016
Poliovirus 1 (no. 80, raw sewage)	0.90	0.029	0.014
Poliovirus 1 (Mahoney, laboratory strain)	0.72	0.029	<0.001
Poliovirus 1 (Sabin, laboratory strain)	0.023	0.004	<0.001
Poliovirus 1 (no. 4, raw sewage)	0.009	<0.001	<0.001
Poliovirus 2 (no. 426, chlorinated water)	0.10	0.033	<0.001
Poliovirus 2 (no. 533, chlorinated water)	0.092	0.020	0.002
Poliovirus 2 (MEF-1, laboratory strain)	0.090	0.010	<0.001
Poliovirus 2 (Sabin, laboratory strain)	0.035	0.006	<0.001
Poliovirus 2 (no. 454, chlorinated water)	0.021	0.003	<0.001
Poliovirus 2 (no. 7, raw sewage)	0.011	<0.001	<0.001
Poliovirus 2 (no. 42, raw sewage)	0.001	<0.001	<0.001
Poliovirus 3 (2 laboratory strains, 3 raw sewage strains, 7 chlorinated water strains)	0.024	0.019	<0.003

Source: Reference 77

The impact of temperature on disinfection efficiency is also significant. For example, Clarke's work in virus destruction by chlorine indicates that contact time must be increased two to three times when the temperature is lowered 10°C.[13] Chlorination requirements for destruction of more newly recognized microbial contaminants have recently been reported by several investigators. For example, recent research regarding treatment necessary to control Giardia has begun to produce detailed results. Disinfection by chlorination can inactivate Giardia cysts, but only under rigorous conditions. Most recently, Hoff et al concluded that (1) these cysts are among the most resistant pathogens known, (2) disinfection at low temperatures is especially difficult, and (3) treatment processes prior to disinfection are important.[14] Their data (shown in Table IV-5) indicate that, "....the resistance of Giardia cysts is....2 orders of magnitude higher than that of the enteroviruses,....and more than 3 orders of magnitude higher than the enteric bacteria...."

Jarroll et al has shown that 99.8 percent of all cysts can be killed by exposure to 2.5 mg/L of chlorine for 10 minutes at 15°C at pH 6, or after 60 minutes at pH 7 or 8.[15] At 5°C, exposure to 2 mg/L of chlorine killed 99.8 percent of all cysts at pH 6 and 7 after 60 minutes. While it took a dosage of 8 mg/L to kill the same percentage of cysts at pH 6 and 7 after 10 minutes, it took a dosage of 8 mg/L to kill all cysts at pH 8 after 30 minutes. These results are illustrated in Figures IV-2 and IV-3.

Hoff et al have collected data from several investigators showing the effects of temperature, pH, and chlorine species on Giardia cyst inactivation.[14] As shown in Table IV-6, inactivation rates using free chlorine decrease with lower temperatures and with higher pH. The results also show that chloramine has a lower disinfection efficiency than free chlorine on Giardia cysts. It is important to note, however, that these tests were conducted with chloramines that were not preformed, meaning that the results are affected by the presence of transient concentrations of free chlorine. Studies by Rubin using only preformed chloramine, shown in Table IV-7, show substantially higher C•T values than the values in Table IV-6, and represent field conditions better than Table IV-6 values.

TABLE IV-5. INACTIVATION OF VARIOUS MICROORGANISMS BY FREE CHLORINE AT 5°C, pH 6.0

Microorganism	Chlorine Conc. (mg/L)	Time* (min)	C·T	Reference
E. coli	0.1	0.4	0.04	14 (16)•
Poliovirus 1	1.0	1.7	1.7	14 (16)
*E. histolytica** cysts	5.0	18	90	17 (17)*
G. lamblia cysts§	1.0	50	50	6 (15)
	2.0	40	80	6 (15)
	4	20	80	6 (15)
	8	9	72	6 (15)
G. lamblia cysts§	2.5	30	75	11 (4)
G. lamblia cysts†	2.5	>60	>150	11 (4)
G. muris cysts	2.5	>60	>150	11 (4)

*For 99% inactivation.
§Cysts from symptomatic carriers.
†Cysts from asymptomatic carriers.
•Reference numbers for this report appear in parentheses.
Source: Reference 14.

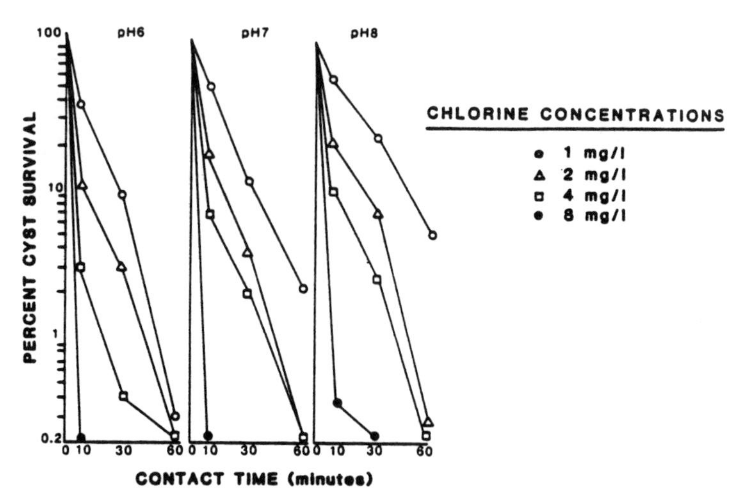

Figure IV-2. INACTIVATION OF G. LAMBLIA CYSTS BY FREE RESIDUAL CHLORINE AT 5°C. (JARROLL et al. [15])

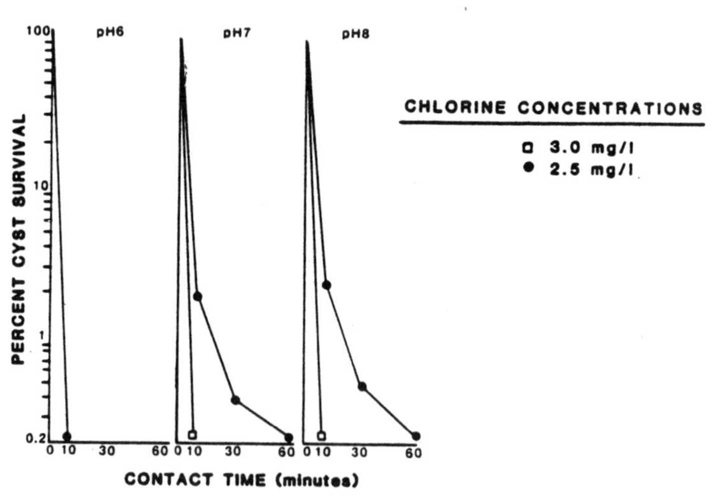

Figure IV-3. INACTIVATION OF G. LAMBLIA CYSTS BY FREE RESIDUAL CHLORINE AT 15°C. (JARROLL et al. [15])

TABLE IV-6. EFFECTS OF TEMPERATURE, pH, AND CHLORINE SPECIES ON GIARDIA CYST INACTIVATION

Form of Chlorine	Temp (°C)	pH	Chlorine Conc. (mg/L)	99% Inactivation Time (min.)	C·T	Reference
Free	25	6	1.5	<10	<15*	6 (15)•
		7	1.5	<10	<15	6 (15)
		8	1.5	<10	<15	6 (15)
Free	15	6	2.5	7	18*	6 (15)
		7	2.5	20	50	6 (15)
		8	2.5	25	63	6 (15)
Free	5	7.8	5	30	150**,§	4 (18)
		7.8	2	180	360	4 (18)
Free	1	7.8	5	60	300**,§	4 (18)
		7.8	2	240	480	4 (18)
Chloramine	18	6.5	2.3	70	161**,§	†
		7.5	2.1	80	168	
Chloramine	10	7.0	1.8	230	414**	†
Chloramine	3	6.5	2.6	180	468**	†
		7.5	2.4	220	528	

*Giardia lamblia cysts.
**Giardia muris cysts.
§Three day old cysts.
†Bingham, A.K. and Meyer, E.A. Unpublished data. Not preformed chloramines.
•Reference numbers for this report appear in parentheses.
Source: Reference 14.

TABLE IV-7. GIARDIA CYST INACTIVATION (99%) USING PREFORMED CHLORAMINES

		Range			
Temp. (°C)	pH	Chloramine Conc. (mg/L)	Time (min.)	C·T	Mean C·T
15	7	5.0-16.6	50-182	825-902	848
15	8	3.2-9.0	58-132	415-525	466

Source: Reference 78

Variations in C·T values for viruses and protozoan cysts, caused by changes in pH and temperature, have been collected and interpreted by Lippy, based on work by others.[19] The values shown in Figure IV-4 for inactivation of viruses by chlorine were developed by White, based on the inactivation of coxsackie A2 viruses.[20] Values for protozoan C·T in the same figure were interpreted from research by White, Jarroll et al and Rice et al.[20,15,4]

In a project intended to evaluate the effect of low chlorine concentrations on Giardia cysts, the U.S. Forest Service conducted tests both (1) with varying residual chlorine concentrations and pH at a constant temperature of 3°C (Table IV-8), and (2) with varying temperature at a constant pH and residual chlorine concentration (Table IV-9). Conclusions of this work were as follows:[21]

1. Initial free chlorine residuals of 0.6 to 1.0 mg/L are effective for Giardia cyst inactivation. However, extensive chlorine contact times are required. Table IV-8 lists recommended contact (exposure) times.

2. Initial chlorine residuals of 0.2 and 0.4 mg/L are inadequate for Giardia cyst deactivation and water system protection unless additional doses are made to maintain these concentrations.

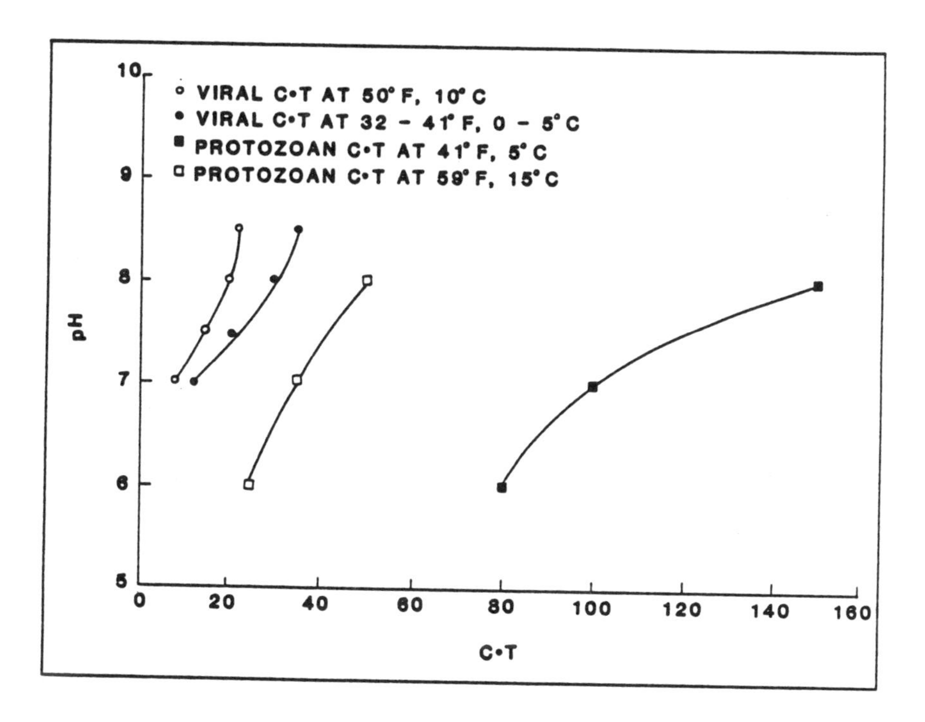

SOURCE: REFERENCE 19

Figure IV-4. C•T RELATIONSHIPS FOR 99-PERCENT INACTIVATION OF PATHOGENIC AGENTS BY CHLORINE

TABLE IV-8. DISINFECTION OF GIARDIA MURIS CYSTS BY FREE CHLORINE

pH	Expected Free Cl^- Residual (mg/L)	Initial Free Cl^- Residual (mg/L)	Final Free Cl^- Residual (mg/L)*	Viability at Time (hours)**							Estimated Minimum Exposure Time (hrs)	Recommended Exposure Time (hrs)
				1	2	3	5	8	12	18		
6.5	1.0	1.06	0.88	0.1	-	0	-	-	-	-	1 - 3	6
	0.8	0.80	0.60	1.7	-	0	-	-	-	-	1 - 3	6
	0.6	0.60	0.36	16.8	-	0	-	-	-	-	1 - 3	6
	0.4	0.42	0.08	45.7	-	0.9	0.2	0	-	-	5 - 8	12
	0.4-boosted	0.39	0.48	-	35.4	2.2	0.2	0	-	-	5 - 8	12
	0.2-boosted	0.24	0.18	-	62.0	44.2	0.9	0.1	0	-	8 - 12	18
7.5	1.0	0.96	0.78	6.5	-	0.5	0	-	-	-	3 - 5	8
	0.8	0.72	0.54	14.6	-	0.8	0.1	0	-	-	5 - 8	12
	0.6	0.60	0.44	56.3	-	3.6	0.2	0	-	-	5 - 8	12
	0.4	0.42	0.08	65.0	-	15.9	0.4	0.2	0	-	8 - 12	18
	0.4-boosted	0.41	0.38	-	42.2	6.4	0.5	0	-	-	5 - 8	12
	0.2-boosted	0.24	0.18	-	69.2	65.4	19.1	5.3	1.4	0	12 - 18	24

*Final free chlorine residuals were measured when the cyst viability reached zero.

**Percentages of viability are relative to controls and are the means of duplicate experiments, a total of at least 500 cysts being counted in each experiment.

Source: Reference 21.

TABLE IV-9. EFFECT OF TEMPERATURE ON FREE CHLORINE DISINFECTION OF GIARDIA MURIS CYSTS

Temperature (°C)	Average Free Chlorine Residual (mg/L)*	Percent Viability At 1 Hour
3°	0.42	44.4
6°	0.39	25.9
12°	0.38	2.2

*Average of free chlorine residuals measured at the beginning and at 1 hour.
Source: Reference 21.

3. The reaction of chlorine on Giardia cysts follows patterns similar to other organisms such as coliform bacteria. The main difference in the exposure time is significantly greater for the Giardia cyst.

Other newly recognized pathogens include the bacteria Campylobacter jejuni, Yersinia enterocolitica, and Legionella pneumophila, as well as viruses such as rotaviruses and Norwalk-like viruses. Blaser et al found that C. jejuni was more susceptible to chlorine disinfection than E. coli under experimental conditions, indicating that effective inactivation of E. coli by chlorination would also be effective for inactivating C. jejuni.[22] Very little information is available from laboratory studies regarding disinfection of Y. enterocolitica. A limited study by Scarpino, however, indicates that Y. enterocolitica is also more sensitive to free residual chlorine than E. coli.[23] L. pneumophilia, while apparently spread primarily by the respiratory route through aerosols rather than by ingestion, does occur and multiply under certain conditions in warm water distribution and plumbing systems, and is capable of growth in tap water.[24] Initial laboratory studies showed viable reductions of at least six orders of magnitude for L. pneumophila almost immediately after dosing with free residual chlorine at 3.3 mg/L.[25] However, subsequent field studies demonstrated that concentrations of chlorine and other biocides that were effective in laboratory studies were not effective in removing L. pneumophila from operating cooling towers.[26,27] More recent laboratory studies have shown conclusively that L. pneumophila grown in unsterilized tap water is much more resistant to disinfection than if grown in culture media.[28] Specifically, at 21°C, a pH range of 7.6 to 8.0, and 0.25 mg of free residual chlorine per liter, 99 percent inactivation of agar-cultured L. pneumophilia required a C•T value of only 2.5, while the same percent inactivation of tap-water-maintained strains required C•T values of 15 to 22.5.[28] One hospital outbreak of Legionella was controlled by maintaining 1 to 2 mg/L of free chlorine at all outlets.[29]

Hoff and Geldreich have summarized the few studies relating to viral agents other than enteroviruses (polioviruses, coxsackieviruses, and echoviruses) as follows: (1) hepatitis A virus may be somewhat more resistant to chlorine than the model enteroviruses, (2) simian rotavirus appears to be 10 times more sensitive to chlorine than poliovirus 1, and (3) no laboratory disinfection studies on Norwalk-like viruses were available as of December, 1982.[30] A more recent study concluded that the Norwalk virus appears to be very resistant to chlorine, specifically, more resistant than poliovirus type 1, human rotavirus, or simian rotavirus.[31] The same study found, for example, that a 3.75 mg/L dose of chlorine was effective against other viruses, but failed to inactivate Norwalk viruses.

Plant-Scale Testing--Within the last 5 years comprehensive testing for virus removal has been conducted on several water treatment plants regarding removal efficiencies of treatment processes including disinfection. While more data on this topic are needed, several conclusions can be drawn from the combined results of available studies, i.e. (1) inactivation or removal of viruses from full-scale treatment plants often do not match inactivation rates achieved in laboratory work, and (2) as indicated in Section II, coliform and turbidity may not accurately reflect the presence of viruses in drinking water, and (3) it is currently uncertain whether a multiple-process plant (conventional treatment) is able to consistently produce an essentially virus-free water, where the source water is heavily contaminated and highly turbid.

Investigation of the efficiency of a water treatment plant in Guadalajara, Mexico, in removing virus, points out (1) the necessity of producing a superior quality water in order to achieve virus inactivation with chlorination, and (2) the additional difficulty added by both a heavily polluted raw water and rain-induced turbidity.[34] Treatment included chemical addition, mixing, flocculation, clarification with pre- and post-chlorination, filtration through rapid sand filters or automatic valveless sand filters, and final chlorination. None of nine finished water samples collected during the dry season contained detectable total coliform bacteria. Seven of the nine samples met turbidity (1.0 NTU), total coliform bacteria (<1 CFU*/100 mL), and total residual chlorine (>0.2 mg/L) standards. Of these seven samples, four contained virus. The minimum turbidity in these four samples was 0.73 NTU and the maximum free chlorine residual was 2.5 mg/L. During the rainy season at this plant, none of 14 finished water samples met turbidity standards, none had a free chlorine residual greater than 1.5 mg/L, and all contained rotaviruses or enteroviruses.[34]

Payment et al described an investigation of seven water treatment plants in the Montreal, Canada, area.[35] Treated water from each plant was sampled twice monthly for 12 months. Three of the plants used postchlorination following filtration; two of these plants also used prechlorination and coagulation plus sedimentation prior to filtration. Analytical results for these three plants are shown in Table IV-10. Of the three plants, Plant 4 and Plant 5 both had heavily virus-contaminated raw water supplies, while Plant 6 was relatively free of viruses. Plant 4 and Plant 5 both had full conventional treatment, although Plant 4 used chlorine dioxide for disinfection. Plant 6 had filtration and postchlorination only. As shown in Table IV-10. Plants 4 and 5 had the lowest mean virus density after disinfection. It should also be noted that the finished water turbidities in all three of these plants at times exceeded an NTU of 1.0.

Stetler et al reported an average virus removal of 98 percent in a water treatment plant they investigated.[36] Viruses removal efficiencies at a treatment plant studied by Gerba et al have previously been tabulated in Section II of this report. The low virus removal (81 percent) at that plant, however, can be partly explained by a number of serious operational deficiencies.

---

*CFU = coliform forming units.

TABLE IV-10. VIRUS REMOVAL AT SEVEN WATER PURIFICATION PLANTS

Water Sample Location	No. Samples Positive/No. Tested (mean virus density, MPNCU*/L) Plant 4	Plant 5	Plant 6
Raw	26/26 (13.31)	24/25 (0.52)	5/25 (0.03)
Filtered	6/27 (0.0021)	0/25 (0.000)	Not Sampled
Finished	1/27 (0.0007)	2/25 (0.0005)	3/25 (0.0014)

*Most probable number of cytopathogenic units.
Source: Reference 35.

In work involving both pilot-scale and full-scale plant testing, O'Connor et al investigated the biocidal efficiency of chlorine, as well as other disinfectants, on bacteria and viruses in water treatment plants on the Missouri River in Missouri.[65] In a test involving inactivation of viruses in raw Missouri River water by chlorine (applied as NaOCl), the investigation found that, "... chlorine at dosages up to 2 mg/L resulted in only marginal inactivation of viruses. Despite the exceptionally adverse conditions for effective disinfection, most of the conventional indicator organisms in raw Missouri River water were killed by chlorination." Later, in pilot-plant studies, using twin process trains including rapid-mix flocculation/sedimentation/dual-media filtration/granular activated carbon, parallel tests were run with and without chlorination preceding the rapid-mix unit. Those tests showed that the average total bacterial removal by the physical process alone, including alum-coagulation, was 90 percent, with most of the microorganism removal (80 to 85 percent) in the sedimentation process. The parallel chlorine treated system achieved 99.9 percent and greater bacterial removal by the end of the sedimentation process.

Summary--

The preceding paragraphs point out that free chlorine is an effective disinfectant against nearly all microbial contaminants, if there are no interfering substances (including turbidity) or circumstances (such as extremes of temperature or pH). Available data also shows, however, that chlorine is relatively more effective, in terms of C·T product for example, against some microorganisms than it is against others. Giardia cysts and viruses are primary examples of more resistant and less resistant organisms for chlorine.

Laboratory data, particularly that by Payment[77] and Liu[12], indicate that chlorination alone can achieve a 99.99 percent inactivation of viruses. When pretreatment and filtration precede disinfection at well operated treatment plants, 99.99 percent inactivation can also be achieved, with 90 to 99 percent removal provided by the pretreatment and filtration processes.[35] In addition, Payment's data or Montreal treatment plants shows also that the combination of prechlorination, sedimentation, and filtration can also inactivate 99.99 percent of viruses.[35]

Chlorine Dioxide

General--

Because of the health concerns with the formation of trihalomethanes in drinking water as a result of chlorination, there has been a widespread intensive search for alternative methods of disinfection. In comparing proposed alternatives to chlorination, there are at least five major items to consider:

- Efficacy of a disinfectant
- Toxic effects of the disinfectant
- Toxic effects of the disinfectant's reaction products
- Maintenance of residual disinfection in the water distribution system
- Cost

Chlorine dioxide (ClO_2) is not widely used as a disinfectant in the United States, though its use for this purpose is relatively common in Europe. Chlorine dioxide cannot be transported because of its instability and explosiveness, so it must be generated at the site of application. The most common method for producing ClO_2 is by chlorination of aqueous sodium chlorite ($NaClO_2$), although the use of sodium chlorate is more efficient. For water treatment, chlorine dioxide is only used in aqueous solutions to avoid potential explosions.

In terms of available chlorine, chlorine dioxide has more than 2.5 times the oxidizing capacity of chlorine. However, its comparative efficiency as a disinfectant varies with a number of factors, as explained in subsequent paragraphs. Chlorine dioxide, when added alone to a water supply, does not produce THM's. As described in the portion of Section II of this report discussing by-products of disinfection, chlorine dioxide can produce by-products which present potential health problems.

Chlorine dioxide and its inorganic reaction products--chlorite and chlorate--may present a higher risk from acute toxicity than chlorine, combined chlorine, or ozone, but its use may be advantageous when the possible generation of carcinogenic by-products of other disinfectants is considered. Chlorine dioxide possesses antithyroid activity. The question of whether chlorine dioxide, chlorite, and chlorate pose a problem of acute toxicity to the segment of the U.S. population that is exceptionally sensitive to agents producing hemolytic anemia, and at what levels, must be addressed.

In practice, the allowable dosage of chlorine dioxide may be constrained by public-health-developed limitations on its by-products, chlorite and chlorate in particular.

Performance--
The performance of chlorine dioxide as a disinfectant has been investigated both by itself and in numerous comparisons with chlorine and other alternative disinfectants. Results and summaries of those studies are presented below.

Laboratory Research and Pilot Studies--Early work by Ridenour and Ingols reported that chlorine dioxide was at least as effective as chlorine against E. coli after 30 minutes at similar residual concentrations.[38] The bactericidal activity of chlorine dioxide was not affected by pH values from 6.0 to 10.0. They also reported that, like chlorine, the efficiency of chlorine dioxide decreased as temperature decreased.

The bactericidal effectiveness of chlorine was compared with that of chlorine dioxide at pH 6.5 and 8.5 in a demand-free buffered system.[39] At pH 6.5, chlorine was more effective. At pH 8.5, chlorine dioxide was dramatically more effective than chlorine. Chlorine dioxide was also significantly more efficient than chlorine in the presence of organic and nitrogenous material. Figure IV-5a shows the concentration-time relationships for 99 percent disinfection of poliovirus 1, coxsackievirus A9 and E. coli by chlorine dioxide.[42] The plots in Figure IV-5b, show that ClO_2, unlike chlorine, is a more effective virucide as pH increases.[42]

Recent work at Ohio State University on inactivation of an amoebic cyst (Naegleria gruberi) indicates that chlorine dioxide is less efficient than ozone, but more efficient than chlorine for cyst inactivation.[40] As shown in Table IV-11, ozone maintained at a constant concentration is definitely superior (based on concentration-time products) at lower temperatures and at pH's of 8 or less. Chlorine dioxide approaches ozone in effectiveness at pH 9 and 25°C, while at pH's above 7, chlorine loses its effectiveness as hypochlorous acid ionizes to the hypochlorite ion. The increasing effectiveness of chlorine dioxide with both pH and temperature identified by other investigators was confirmed for N. gruberi cysts in this work. The authors also demonstrated that, based on concentration-time products, N. gruberi is five times as resistant to inactivation by chlorine dioxide as is poliovirus 1, and 20 times more resistant than E. coli.[40]

Symons et al studied both the effectiveness of alternative disinfectants on microbial contaminants and methods of controlling THM's.[41] In that report, they combined the results of other researchers to provide graphic comparisons of effectiveness. Figures IV-6 and IV-7 present the results of those studies. The relative efficiencies of the various disinfectants in these studies varied both according to target organism (E. coli and poliovirus 1), and with changes in pH and temperature. Figure IV-7 shows that chlorine dioxide at pH 7 and HOCl at pH 6, produced similar inactivation rates for E. coli and poliovirus 1. Hypochlorite ion (OCl^-), monochloramine, and dichloramine were all less effective.

Although several utilities have experimented with chlorine dioxide as an alternative disinfectant for THM control, Mobile, Alabama, is the only utility serving more than 75,000 in 1983, where permanent plant modifications have been made. In Mobile, Alabama, chlorine dioxide is used year round, with chlorine dioxide addition at the raw water pumping station and following treatment, a combination

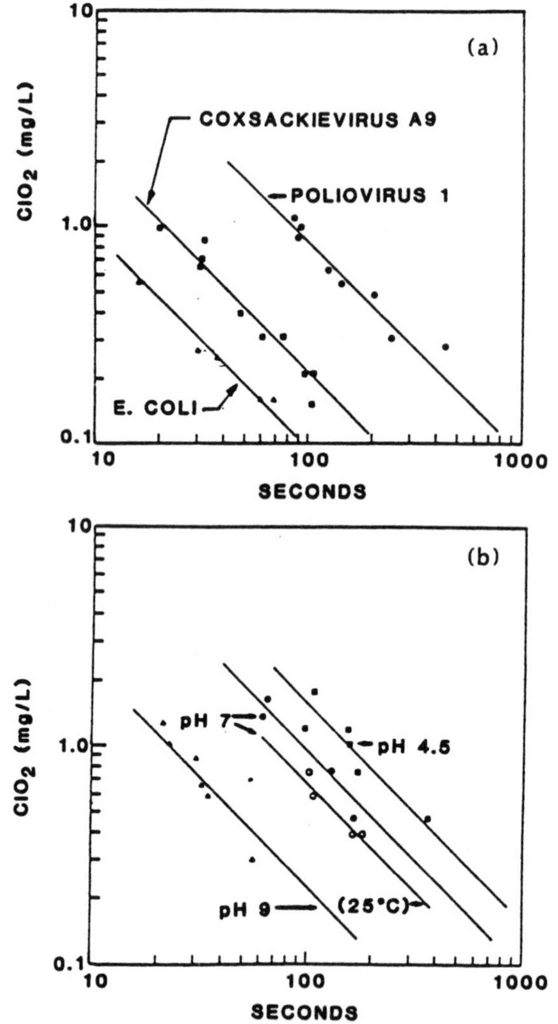

Figure IV-5. (a) CONCENTRATION-TIME RELATIONSHIP FOR 99% DESTRUCTION OR INACTIVATION OF POLIOVIRUS 1, COXSACKIEVIRUS A9 AND E. COLI BY CHLORINE DIOXIDE AT 15°C AND pH 7.0.
(b) EFFECT OF pH ON INACTIVATION OF POLIOVIRUS 1 AT 21°C AND pH 4.5, 7 AND 9, AND AT 25°C AND pH 7. (COURTESY ANN ARBOR SCIENCE PUBLISHERS. [42])

TABLE IV-11. COMPARISON OF C·T PRODUCTS FOR THE INACTIVATION OF N. GRUBERI CYSTS WITH DIFFERENT DISINFECTANTS

Disinfectant	Temperature °C	C·T Product					Reference
		pH 5	pH 6	pH 7	pH 8	pH 9	
Chlorine Dioxide	5			15.47			This work[40]
	15			9.48			
	25	6.35	5.25	5.51	3.92	2.91	
	30	4.25	4.12	3.77	2.95	2.29	
Chlorine	25	7.72		12.1		197	Rubin et al[43]
Ozone	5			4.23			Wickramanayake et al[44]
	15			2.04			
	25	1.33	1.55	1.29	1.35	2.12	
	30			1.21			

Source: Reference 40.

of chlorine dioxide and free chlorine is added before pumping finished water to the distribution system. Table IV-12 shows the results of THM sampling before and after the plant's conversion to the use of chlorine dioxide.

The data indicate that although some THM's are still formed, the use of chlorine dioxide clearly reduced THM formation relative to use of free chlorine only. With the use of chlorine dioxide it is possible to maintain a disinfectant residual in the distribution system.

Summary--

Chlorine dioxide is an effective disinfectant, which has been shown to be equal to or greater than chlorine in its ability to inactivate viruses and Giardia. It can, therefore, provide equal or greater percent reductions than chlorine. Its use to date, has been limited by its higher cost in comparison to chlorine, and the technical knowledge required to operate and maintain on-site generation equipment.

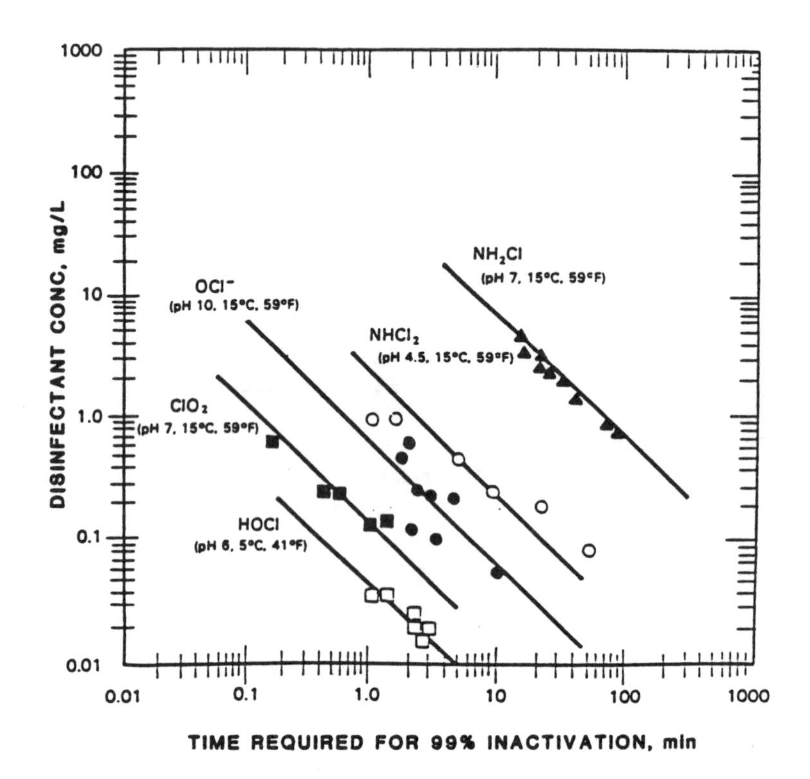

Figure IV-6. INACTIVATION OF E. COLI (ATCC 11229) (SOURCE: REF. 41)

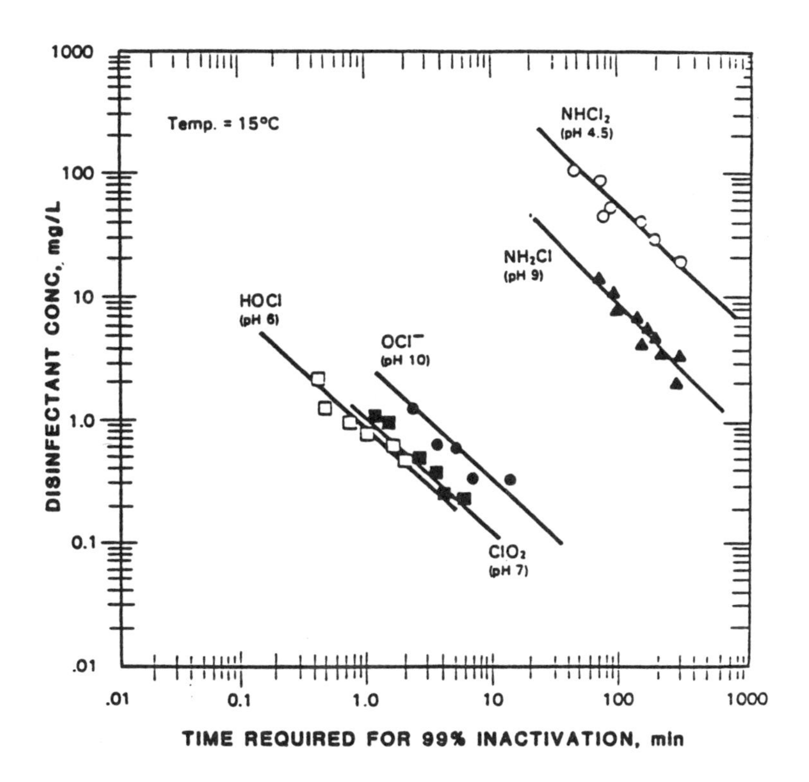

Figure IV-7. INACTIVATION OF POLIOVIRUS 1 (MAHONEY) (SOURCE: REF. 41)

TABLE IV-12. AVERAGE OR RANGE IN INSTANTANEOUS THM CONCENTRATIONS AT MOBILE, ALABAMA, USING CHLORINE DIOXIDE AND FREE CHLORINE DISINFECTION

Sample Location	Before Conversion, Free Chlorine Disinfection	After Conversion, Chlorine Dioxide Disinfection
Plant product water	140 μg/L	40 μg/L
Distribution system-average	200 μg/L	40-90 μg/L

Chloramination

General--

In aqueous systems, chlorine reacts with ammonia (NH_3) to form chloramines as follows:

$$NH_3 + HOCl \rightleftarrows NH_2Cl + H_2O$$
monochloramine

$$NH_2Cl + HOCl \rightleftarrows NHCl_2 + H_2O$$
dichloramine

$$NHCl_2 + HOCl \rightleftarrows NCl_3 + H_2O$$
trichloramine (nitrogen trichloride)

Chloramines have many properties different than HOCl and OCl^- forms of chlorine. They exist in various proportions depending on the relative rates of formation of monochloramine and dichloramine, which change with the relative concentrations of chlorine and ammonia, with pH, and with temperature. Above about pH 9, monochloramines exist almost exclusively; at about pH 6.5, monochloramines and dichloramines coexist in approximately equal amounts; and below pH 6.5 dichloramines predominate, while trichloramines exist below about pH 4.5.

Chloramines are generally less effective than chlorine for inactivating bacteria, viruses, and protozoans at equal dosages and contact times. At temperatures below 5°C, however, some data indicate that chloramines which are not preformed may be nearly as effective as free chlorine in disinfecting Giardia (see Table IV-6). At higher temperatures, and for other microbial contaminants, the lower efficiency of chloramines can be overcome by the use of higher dosages and longer contact times where feasible. As indicated in Tables IV-6 and IV-7, however, it is important to note that there is a significant difference in effectiveness between preformed chloramines and those that are not preformed. When non-preformed chloramines are used, transient concentrations of free chlorine are present, which increases the disinfection potency in comparison to the use of preformed chloramines. In practice, as described later, ammonia and chlorine are usually added separately in water treatment plants, so that data on preformed chloramines may not be relevant to field conditions.

Chloramines have also been demonstrated to be toxic in some situations, specifically to patients on artificial kidney machines undergoing hemodialysis.[45] Chloramines have been shown to be capable of oxidizing hemoglobin to methemoglobin in vitro. This property of NH_2Cl has been responsible for producing hemolytic anemia in patients undergoing dialysis.[45] However, no indication of hemolytic anemia has been found in animals consuming NH_2Cl orally at concentrations as high as 100 mg/L.[46] There is also a lack of information about the carcinogenicity of chloramine by-products. Monochloramine possesses mutagenic properties in bacteria and produces changes in the livers of mice which are commonly associated with carcinogens. The importance of these properties, however, must await the completion of lifetime studies with experimental animals.

On the other hand, chloramines possess greater stability of residuals than chlorine, they generally reduce taste and odor problems caused by chlorine, and produce fewer trihalomethanes. Chloramines are the least costly of the major alternative drinking water disinfectants.

Performance--

The combination of chlorine and ammonia produces a more stable disinfecting residual than produced by chlorine alone. However, the germicidal efficiency of chloramines is substantially less than hypochlorous acid, particularly with respect to viruses.

A detailed history of the discovery and use of chloramines has been presented by Wolfe et al.[47] The authors note that prereacted chloramines were first used in the United States in 1917 by the Denver Union Water Company, Denver, Colorado, to prevent aftergrowth problems. Wolfe et al also describe the apparent discrepancy between laboratory tests and field tests regarding the effectiveness of chloramines for inactivating coliforms, indicating that numerous field tests have shown chloramines to be as effective as chlorine on coliforms when sufficient contact times are provided.[47] One explanation of the discrepancy between field and laboratory effectiveness tests, as noted earlier, is that in laboratory work the chloramines are preformed and the microorganisms are added afterward for testing, while conventional practice for chloramination in the field is to add ammonia to the water first, and chlorine later.[48] This practice, under some conditions, could result in the presence of free chlorine for several minutes with consequent rapid inactivation of microorganisms. Conditions under which this could happen are at lower temperatures and pH values above or below optimum (pH 8.3). Therefore, the much slower inactivation rates shown by preformed chloramines may not be directly relevant to chloramination in the field.[48]

The relative ineffectiveness of chloramines, and particularly preformed chloramines, for inactivating both Giardia and viruses is shown in Tables IV-1, IV-2, IV-6, and IV-7. As indicated in those tables, chloramines can inactivate these organisms, but are much less efficient than chlorine, chlorine dioxide, or ozone in doing so.

Table IV-13 summarizes data on virus inactivation rates at varying pH at 10°C in filtered secondary sewage effluent. The constants in Table IV-13 were determined from experimental results using model f_2 virus which was seeded in filtered secondary sewage effluent, to simulate field contamination conditions, buffered to the desired pH value. Chlorine solution was flash mixed at a dose of 30 mg/L and the time of persistence of any free chlorine as well as virus survival was determined.[8] Data in Table IV-13 do not, however, show a simple pH effect. Instead, they show the effect of pH on the reaction of free chlorine with ammonia

TABLE IV-13. VIRAL INACTIVATION RATE CONSTANTS AT VARYING pH OF SEWAGE AT 10°C

Sewage pH	Flash Mixing of 30 mg/L Chlorine: Viral Inactivation Rate Constant (min^{-1})	Flash Mixing of 30 mg/L Chlorine: Free Chlorine Duration (sec)	Viral Kill (percent)
5.0	120.0	110.0	>99.9999
5.5	72.0	24.0	>99.9999
6.0	44.0	10.0	99.93
6.5	26.0	<5	92.0
7.0	16.0	<5	76.0
7.5	9.5	<5	55.0
8.0	5.6	<5	40.0
8.5	3.5	<4	25.0*
9.0	2.0	<4	15.0*
9.5	1.2	<4	10.0*

*Essentially control survival.
Source: Reference 8.

nitrogen to form chloramines. At neutral or higher pH, the reaction of free chlorine with ammonia is rapid and combined chlorine, a poor virucide, is formed. At lower pH values, the reaction is slow, and unreacted free chlorine is available for sufficient time to inactivate the viruses.

The need for applying chlorine to distribution systems is recognized as necessary to control regrowth of coliform organisms and control growth of organisms which cause taste, odor, or corrosion. Whether the chlorine should be applied as free chlorine or combined chlorine (chloramine) has always been a matter of controversy. White has presented case histories of both methods of application to distribution systems, dating back to the 1930's.[20] He also describes proponents of the free residual method as "...usually those confronted with a serious situation that needs drastic action and...quick to admit serious consumer complaints while the system is being cleaned." Proponents of the combined available residual method, on the other hand, make the following points:

"1. The combined residual, while less potent, is more persistent and will eventually penetrate farther for a longer time.

2. If a free residual is pushed forward and the system is badly contaminated, this free residual in its travel will gradually be turned back toward the dip and then to the hump of the breakpoint curve and will eventually become all combined available residual. Further, during this process whereby the free residual chlorine is being converted to combined residual, tastes will result from the inevitable formation of nitrogen trichloride and dichloramines.

3. This eventual formation of combined residual chlorine is certain to occur in the consumer service pipes long after the distribution pipes have been cleaned."[20]

At the time this document is being prepared, research is still necessary to define the conditions and parameters which should be used to determine whether free residual chlorine or combined residual chlorine is preferable in a particular distribution system.

Because laboratory inactivation efficiencies of chloramines have already been presented earlier in this section, the following paragraphs will describe pilot-scale and full-system experiences through case histories.

Case Histories--

Ten Utility Survey--Many utilities have chosen alternative methods of disinfection as a means of controlling THM's. Among the methods available, disinfection using chloramines has become the most widely practiced. The ammonia application points, however, vary from one plant to another. Table IV-14 lists ten utilities which modified their treatment facilities for disinfection using combined chlorine.[49] The points of chlorine and ammonia application for each plant are compared in Table IV-14. With the exception of Brownsville and Houston, Texas, where raw water is ammoniated prior to chlorine addition to ensure adequate ammonia mixing, all other facilities add ammonia at some point following chlorination. As illustrated in Table IV-14, most utilities add chlorine just before filtration (Louisville, Kentucky, applies chlorine slightly earlier, immediately following coagulation, while Topeka, Kansas, uses prechlorination to the extent possible while still meeting regulations). Several facilities also apply chlorine at a second point in the treatment process, usually after sand filtration.

TABLE IV-14. CHLORINE AND AMMONIA APPLICATION POINTS OF UTILITIES USING CHLORAMINE DISINFECTION

Utility	Population	Chlorine Application Points	NH_3 Application Points
Kansas City, KS	190,000	After sand filters	Following chlorine addition
Louisville, KY	333,000	After coagulation After sand filers	10 minutes after each application of chlorine
Contra Costa, CA	150,000	Before sand filters	After sand filters
Miami, FL	1,500,000	Before sand filters	30 - 60 Sec. after chlorination, before sand filters
Occoquan, VA	700,000	Before sand filters	After sand filters
Tampa, FL	455,000	Before sand filters After sand filters	After sand filters
San Diego, CA	850,000	Before sand filters	1 mg/l free Cl_2 residual maintained through filters before NH_3 addition
Brownsville, TX	85,000	Before mixing chamber After sand filters Before distribution	Raw water
Houston, TX	2,000,000	After raw water pump and at clearwell before distribution	Raw water
Topeka, KS	180,000	After presedimentation and before rapid mix. Also before sand filters	Rapid mix and before filters

Source: Reference 49.

Of the ten utilities listed in Table IV-14, five add chlorine before sand filtration and ammonia after sand filtration, thereby maintaining a free chlorine residual through the filters. The remaining five utilities listed in Table IV-14 operate with combined chlorine formation prior to filtration.

Some plants operating without free chlorine prior to the filters have experienced problems during summer months due to algae growths in the walls of both settling basins and filters. They have also had difficulty with algae carrying over from settling basins to filters. The City of Brownsville, Texas, attempted for a short time to eliminate prechlorination at the mixing chamber (before ammonia feed facilities were on-line), but found it impractical due to excessive algae growth in clarifiers and filters. Consequently, prechlorination is still practiced at Brownsville.

With the exception of Kansas City and Topeka, Kansas, all of the utilities using chloramine disinfection operate their THM control facilities on a year-round basis, and none reported a noticeable drop in bacteriological quality of the finished water, based generally on coliform and standard plate count testing.[49] The results of water quality sampling before and after introduction of chloramines into several water supply systems indicate that certain quality characteristics actually are enhanced by the use of chloramines. For example, the City of San Diego, California, found it unnecessary to rechlorinate standpipes and various locations within the distribution system after conversion to chloramine disinfection. Also, standard plate counts proved to be as low as, if not lower than, with free chlorine, and city residents have reported an improvement in the taste of the water. Contra Costa County, California, reported a definite improvement in turbidity of the finished water both at the plant and in the distribution system. Miami, Florida, was the only utility reporting adverse effects from the use of chloramines. Although bacteriological quality of the treated water remained unchanged, the utility received complaints about color. When Miami previously operated with breakpoint chlorination, treated water typically contained 3 to 5 color units. After conversion to chloramines, the plant produced a water with an average of 13 color units.[49]

Philadelphia Suburban Water Company (PSWC)--Chloramines have been used at the PSWC's Pickering Creek water treatment plant since 1930, and has helped to reduce coliform counts in the raw water from values as high as 100,000 organisms/mL to nearly zero in the finished water.[50] PSWC serves more than 850,000 people in 52 communities west of Philadelphia, and the Pickering Creek plant is one of three treatment facilities serving the area. Chloramines are applied exclusively for disinfection, as part of the pretreatment process, followed by flash mixing, flocculation, coagulation, clarification, rapid-sand filtration, and postchlorination. Chloramination is accomplished at the plant by applying aqua ammonia first and chlorine later, usually in a 1:3 (NH:Cl) ratio. Despite the fact that total bacteria and coliform counts in the raw water have increased by one to two orders of magnitude since 1930, coliform counts in water at the top of the filter have not changed in that same time period.

Fort Meade, Maryland--The street mains and internal plumbing for two barracks at Fort George G. Meade, Maryland, were used as a test water distribution system to evaluate the stability and effectiveness of three disinfectants.[51] Free chlorine, combined chlorine, and chlorine dioxide were tested at 0.2 mg/L and 1.0 mg/L as residual available chlorine. Among the conclusions for the study were the findings that: 1) free chlorine was the most effective residual disinfectant, 2) combined chlorine was the most stable in the pipe network, and 3) the half-lives of the three disinfectants in static tests were 140 minutes, 93 minutes, and 1,680 minutes, respectively, for free chlorine, chlorine dioxide, and combined chlorine.

Summary--

As the preceding pages indicate, chloramination is capable of the same percent inactivations of microorganisms, but is significantly less efficient in doing so, in terms of C•T product comparisons. Chloramines are, however, the least costly of the major alternatives to chlorination, and produce fewer trihalomethanes. Chloramines can be toxic in certain limited situations, and the carcinogenicity of chlorine byproducts is unknown at the present time.

Ozonation

General--

Ozone is the most potent and effective germicide used in water treatment. Only free residual chlorine can approximate it in bactericidal power, but ozone is far more effective than chlorine against both viruses and cysts.

The City of Nice, France, has used ozone for disinfection of municipal drinking water since 1906.[52] It has been reported that nearly a thousand cities in the world purify their water with ozone, with the largest existing plant having a capacity of 238 mgd. A new treatment plant in Los Angeles, California, is under construction which will be the largest using ozone (though not primarily for disinfection), at a capacity of 580 mgd. That plant, along with two more known to be under construction, three in various phases of design or bidding, and more than 20 in operation in the U.S. are listed in Table IV-15. As indicated in the table, the primary purpose for the use of ozone in the U.S. now is usually for something other than disinfection.

The advantages of using ozone are its high disinfection effectiveness; its ability to ameliorate many problems of odor, taste, and color in water supplies; and the fact that upon decomposition, it apparently does not form THM's or other carcinogenic by-products. Like other oxidants, however, it can cause incomplete oxidation of some organic compounds, and it can form precursors of carcinogenic compounds. In addition, its potency is relatively unaffected by pH level or ammonia content. It can also be used effectively for oxidation of many organic compounds, and its oxidative strength also causes changes in the surface charges of both dissolved organic compounds and colloidal-sized particles. The effects of these surface charge changes may lead to what is known as microflocculation of dissolved organics, and coagulation of colloidal particles. In both cases, the use of ozone reduces waste disposal volumes, concerns, and costs. The combination of effects caused by ozone means that, under the right conditions, the use of ozone for one purpose will also result in other beneficial reactions. For example, even if ozone is applied primarily for iron and manganese oxidation, it will provide significant disinfection at that point in the treatment process, as well as oxidation of organics and microflocculation. The microflocculation and improved coagulation brought about by ozone also result in longer filter runs in many cases. Finally, the oxygenated organic materials produced by ozone are more readily biodegradable.

Nonetheless, the use of ozone does have disadvantages. In aqueous solution, ozone is relatively unstable (half-life of 165 minutes in distilled water at 20°C) and as noted above, is highly reactive. Therefore, it does not provide a long-lasting

TABLE IV-15. U.S. POTABLE WATER TREATMENT PLANTS USING OZONE, 1984

Location	Primary Purpose of Ozone	Startup Date	Av. Flow Rate mgd	Av. Flow Rate m^3/day
IN OPERATION				
Whiting, IN	T & O[+]	1940	4	15,142
Strasburg, PA	disinf.	1973	0.1	379
Grandin, ND	Fe & Mn	1978	0.05	189
Saratoga, WY	T & O	1978	3.5	13,249
Bay City, MI	T & O	1978	40	162,487
Monroe, MI	T & O	1979	18	73,119
Newport, DE	disinf.	1979	0.25	1,016
Newport, RI	THM precursors; T&O; color	1980	5	20,311
No. Tarrytown, NY	T & O	1980	1.2	4,875
Kennewick, WA	color;T&O	1980	3	12,187
Elizabeth City, NC	color	1981	5	20,311
Casper, WY	disinf.	1982	5	20,311
Ephrata Borough, PA	T & O	1982	0.145	549
New Ulm, MN	Fe & Mn	1982	2.6	10,562
South Bay, FL	color	1982	2.2	8,937
Rockwood, TN*	flocculation; THM precursors	1982	6	22,712
Lakeport, CA	organics**	1982	<2	<10,000
Mills Wardell, WY	disinf. & T & O	1983	4	15,142
Potsdam, NY	color	1984	1	3,786
Beria, OH*	THM & T&O	1984	3.6	14,348
Belle Glade, FL*	color, THM precursors & algae	1984	6	22,712
Stillwater, OK	color	1984	5	18,927
Spirit Lake, IA	T & O	1984	3	11,358

TABLE IV-15 (Continued)

Location	Primary Purpose of Ozone	Anticipated Startup or Bid Date	Av. Flow Rate mgd	m^3/day
UNDER CONSTRUCTION		Startup		
New York, NY	organics	1984	3 (pilot)	11,358
Hackensack, NJ	color, Fe & Mn; THM precursors	1986	100	378,540
Los Angeles, CA	microfloc-ulation & organics	1986	580	2,195,531
UNDER DESIGN		Bid		
Myrtle Beach, SC*	color; THM precursors	1987	30	121,865
Clinton, IL	Fe	1985	1.5	5,679
Edmond, OK	color	1984	12	45,424
PILOT PLANT STUDIES				
Rocky Mount, NC	color;THMs			
Celina, OH	T&O; THM precursors			

Source: Reference 53

*2-stage ozonation plants.
**Ozonation prior to GAC adsorption (BAC process).
+Tastes and odors.

disinfecting residual in drinking water. Capital costs of ozone installations are relatively high. Because it must be produced electrically on-site as it is needed and cannot be stored, it is difficult to adjust treatment to variations in load or to changes in raw water quality with regard to ozone demand. As a result, ozone historically has been found most useful for supplies with low or constant demand, such as groundwater sources. Moreover, although ozone is a highly potent oxidant, it is selective, and by no means universal in its action. In some instances, ozonation of river waters heavily polluted with organic matter can fragment large organic molecules into substances more easily metabolized by microorganisms. This fragmentation, coupled with the inability of ozone to maintain an active residual in the distribution system, can lead to increased slime growths and consequent deterioration of water quality during distribution.

In many ways, the desirable properties of ozone and chlorine as disinfectants are complementary. Ozone provides fast-acting germicidal and virucidal potency, commonly with beneficial results regarding taste, odor, and color. Chlorine provides sustained, flexible, controllable germicidal action that continues to be beneficial during distribution. Thus, it would seem that in the right circumstances a combination of ozonation and chlorination might provide an almost ideal form of water supply disinfection. The procedure is being employed in a number of places, most particularly in the Netherlands at Amsterdam on Rhine River water. Ozone is used initially for phenol oxidation, viral destruction, and general improvement in physical quality; after normal water treatment, postchlorination is used to assure hygienic quality throughout the system. It appears possible that this approach will find increasing use in the U.S. as improvements in ozone-generating equipment are made.

In other circumstances, where chlorine may have detrimental effects even when combined with ozone, the use of ozone could be combined with other alternative disinfectants. Under the right conditions, for example, ozone and chloramination may prove very effective.

Even in cases where disinfectants are used in sequential combination, however, it is important to note that they would still not be effective if applied to a water that is turbid or has a relatively high total organic carbon concentration.

Performance--

Laboratory Research and Pilot Plant Studies--Ozone has been found to be many times more effective than chlorine in inactivating polioviruses. Under experimental conditions, an identical dilution of the same strain and pool of virus, when exposed to chlorine in residual amounts of 0.5 to 1.0 mg/L, and to ozone in residual amounts of 0.05 to 0.45 mg/L, was inactivated within 2 minutes by ozone, while 1 1/2 to 2 hours were required for inactivation by chlorine.

Recent comparisons using the C T (concentration•time) product have placed a more precise definition on the inactivation efficiencies of ozone and chlorine. As shown in Tables IV-1, IV-2, and IV-16, ozone and chlorine are very similar in inactivation efficiency with respect to E. coli, but ozone is one order of magnitude more efficient against poliovirus 1, and two to three orders of magnitude more efficient against Giardia lamblia cysts. Hoff et al also described work by several investigators showing that ozone appears to be more efficient than chlorine for inactivation of Giardia.[14]

Ozone has also been reported to be several times faster than chlorine in germicidal effects on Entamoeba histolytica. For most cases, an ozone residual of 0.1 mg/L for 5 minutes is adequate to disinfect a water low in organics and free of suspended material. Organic material does exert a significant ozone demand.

Tests regarding the effect of pH changes on ozone disinfection efficiency indicate that the magnitude and type of effect is specific for individual organisms. Generally, however, there seems to be relatively little change in resistance to inactivation for many organisms in the pH range of 5 to about 8.[44] Giardia muris cysts, for example, were found to be slightly more sensitive to ozone at pH 5 than at 7, but was nearly one-and-a-half times less resistant at pH 9.[44] Kessel et al, on the other hand, reported that the rate of inactivation of E. histolytica cysts with ozone did not vary significantly in the pH range of 6 to 9, but increased at pH 5.[54]

Pilot testing in a 10-gpm system over a 10-month period (October 1986 to August 1987) documented several of the beneficial reactions which can result from the

TABLE IV-16. COMPARISON OF 99 PERCENT INACTIVATION OF VARIOUS MICROORGANISMS BY CHLORINE AND OZONE

Disinfectant	Temp. °C	pH	C·T* mg·min·L-1	Reference
Chlorine				
E. coli	5	6	0.04	7 (16)•
Poliovirus 1	5	6	1.7	7 (16)
Giardia lamblia cysts	5	6	75	8 (4)
Ozone				
E. coli	1	7.2	0.02	9 (80)
Poliovirus 1	5	7.2	0.22	10 (81)
Giardia lamblia cysts	5	7	0.53	11 (6)

*C·T = Concentration of disinfectant in mg/L multiplied by contact time in minutes for 99% inactivation.
•Reference numbers for this report appear in parentheses.
Source: Reference 73.

use of ozone.[55] Joost et al applied a range of ozone doses during the test and noted the following results with a 3.0 mg/l ozone dose: (a) it consistently lowered filtered water turbidity from 0.24 NTU (without ozone) to 0.14 NTU over a 5-month operation period; (b) it also allowed a reduction in alum coagulant dose from 9 mg/l to 4 mg/l in order to achieve equivalent finished water turbidity; (c) finished water particle counts were an order of magnitude higher for ozonated versus non-ozonated waters. A 1.5 mg/l ozone dose before coagulant addition resulted in about 85 percent of the turbidity reduction attained with 3.0 mg/l, while other process enhancement showed similar behavior with the lower ozone dose. One other conclusion of the study was that while treated water TTHM concentrations with ozonation prior to a 2.4 mg/l chlorine dose were about one-half those without ozonation, a distribution system TTHM goal of 0.020 mg/l could only be met by using ozone and either chloramines or a chlorine dose less than 2.4 mg/l.[55]

Plant-Scale Results--The City of Paris, France, now ozonates its water to such a degree that a minimum of 0.4 mg/L residual ozone is maintained for a 15 minute contact. At this ozone concentration, hygienic security, including destruction of viruses, has been found equal to that provided by chlorination.

The City of Monroe, Michigan, uses ozone for taste and odor control in their 18 mgd water treatment plant.[56] Ozone is applied to either raw or settled water at dosages which vary seasonally from 0.5 to 1.5 mg/L. Ozone is introduced through porous aluminum oxide diffusers into a two-compartment, 10-minute detention contact chamber. Effective control of taste and odor was achieved as evidenced by a lack of complaints received by the City. Application of ozone provided secondary benefits including reduction in chlorine demand, partial to complete disinfection, improved alum coagulation and settling. The THM formation potential was reduced by 40 to 50 percent.

The City of Lakeport, California, installed water treatment facilities in 1983 to process water from Clear Lake, containing, extremely high concentrations of the taste- and odor-causing algae (Aphanizomenon, Oocystis, Melosira, Anabaena, and Microcystis). Ozone in addition to postfilter, granular activated contactors were chosen to supplement coagulation, flocculation, sedimentation and filtration.[57] The 500 gpm plant includes a 24 pounds per day ozonator. Ozone can be applied either to raw or filtered water.

Operation during the first year at Lakeport produced a high quality water and there were virtually no consumer taste and odor complaints. Samples of raw water, clarified water, filtered effluent and carbon column effluent were collected on one occasion and analyzed for reduction in trihalomethane formation potential. Ozone at an approximate dosage of 1.5 to 2.0 mg/L was added prior to flocculation and after filtration. Ozonation combined with clarification and filtration provided 85 percent reduction in trihalomethane formation potential from 89 μg/L to 13 μg/L. Polishing through the granular activated carbon column provided an additional 4 percent removal to an overall plant removal of 89 percent.

Summary--

Ozone is the most potent and effective germicide in water treatment. As shown in Tables IV-1, IV-2, IV-11, and IV-16, ozone can easily provide 99.9 percent inactivation of Giardia cysts, a statement which cannot be made about chlorine or chloramines, i.e., the C•T products for ozone are very small in comparison to the other two disinfectants. In addition, it is more efficient than chlorine for inactivating viruses, and is roughly equal in inactivation efficiency to chlorine dioxide with respect to both Giardia cysts and viruses.

Use of ozone for water disinfection is not common in the United States, due to a combination of first cost (equipment cost), and a more complex technology than chlorine. It apparently does not form THM's, although its reactions can form precursors of carcinogenic compounds. Due to its relative instability and high reactivity, however, it does not provide a long-lasting disinfecting residual in drinking water, and therefore would have to be used with one of the other major disinfectants to provide such a residual.

OTHER APPLICABLE TECHNOLOGIES

The following is a list of Other Applicable Technologies:

- Iodine
- Bromine
- Ultraviolet radiation
- Heat treatment

Iodine

General--

Iodine possesses the highest atomic weight of the four halogens and is the least soluble in water, with a solubility in water of 339 mg/L. It has the lowest oxidation potential, and reacts least readily with organic compounds. Taken collectively, the characteristics mean that low iodine residuals should be more stable and, therefore, persist longer in the presence of organic (or other oxidizable) material than corresponding residuals of any of the other halogens.

Iodine reacts with water in the following manner:

$$I_2 + H_2O \rightleftarrows HOI + H^+ + I^-$$

Diatomic iodine (I_2) reacts with water to form hypoiodous acid (HOI) and iodide ion (I^-). At pH 5.0 approximately 99 percent of the iodine is in the form of I_2. As the pH increases from 6.0 to 8.0, the percent of iodine as I_2 decreases, while the percent of iodine as HOI increases. At pH 7.0, the two forms are present in about equal concentrations, while at pH 8.0, only 12 percent is present as I_2 and 88 percent is present as HOI.

Hypoiodous acid, in turn, can be converted to hypoiodite ion (OI^-) as follows:

$$HOI \rightleftarrows H^+ + OI^-$$

At pH values over 4.0, the hypoiodous acid undergoes dissociation, as shown above. At pH values over 8.0, HOI is unstable and will not form the hypoiodite ion but will decompose according to the following equation:

$$3HOI + 2(OH^-) \rightleftarrows HIO_3 + 2H_2O + 2I^-$$

This results in the formation of iodide (I^-) and iodate (IO_3^-).

Some concern has arisen over the possible harmful effects of iodinated water on thyroid function, on possible sensitivities to iodine itself, and objectionable tastes and odors. During a policy review in 1982, EPA's Criteria and Standards Division, Office of Drinking Water, confirmed its recommendation that iodine disinfection is acceptable for short-term, limited or emergency use, but is not recommended for long-term or routine community water supply. The policy confirmation was based both on additional information regarding adverse health reactions and on current data on average daily intakes of iodine versus the Recommended Dietary Allowance (RDA) for intake of iodine by adults.[58] The RDA ranges from

0.08 to 0.14 mg/day. The average daily intake of iodine, based on dietary studies, appears to be at least twice the RDA, ranging from 0.24 to 0.74 mg/day. Based on a National Academy of Sciences recommendation that iodine intake be reduced to approximately the RDA, EPA policy recommends that the use of iodized salt be discouraged in areas where iodine is used as a drinking water disinfectant.

Regarding taste and odor problems, few subjects, if any, are able to detect by either color, taste or odor a concentration of 1.0 mg/L elemental iodine in water. Iodide ion cannot be detected in concentrations exceeding 5.0 mg/L. When the concentration of iodine is between 1.5 and 2.0 mg/L, many people will be able to detect a taste which they may not consider to be objectionable.[59]

Since iodine does not combine with ammonia or many other organic compounds, the production of tastes and odors is minimized. Iodine does not produce any detectable taste or odor in water containing several ppb phenol, whereas chlorine produces highly objectionable taste and odor under similar conditions.

Performance--

Laboratory Research--Research into the bactericidal effects of iodine has shown only a slight difference in effectiveness between pH 5.0 and pH 7.0. At pH values of 9 or above, a marked reduction in efficiency occurs. Apparently, hypoiodous acid formed at higher pH values is less bactericidal than free iodine. At pH 7.0, it has been found that 99.99 percent kill of the most resistant of six bacterial strains tested could be achieved in 5 minutes with a free iodine residual of 1.0 mg/L.[60]

In general, it has been found that the bactericidal action of iodine resembled that of chlorine with respect to temperature and pH, but higher concentrations of iodine were required to produce comparable kills under similar conditions. However, toward higher organisms and especially toward cysts and spores, iodine and bromine may have some advantages.[61] Rapid (5 to 10 minutes) destruction of cysts with iodine concentrations of 5 to 10 mg/L have been reported for several different cysts in different neutral waters.

The relative efficiencies of hypochlorous acid and elemental iodine is demonstrated by the fact that iodine takes 200 times longer to produce the same amount of destruction of two types of coxsackieviruses.[13]

HOI (hypoiodous acid), which results from the hydrolysis of I_2 as the pH of solution increases above 6, destroys viruses at a rate considerably faster than that achieved by I_2. Iodate and iodide are essentially inert as disinfectants.

In summary, iodine's advantage over chlorine is that it does not react with ammonia or similar nitrogenous compounds. As a result, the bactericidal and cysticidal potency of I_2 and the virucidal effectiveness of HOI are not adversely affected by the presence of such nitrogenous pollutants, whereas the action of a given applied concentration of chlorine is affected either by exertion of demand or formation of chloramines. Moreover, by appropriate control of dosage, pH, and other manageable conditions, aqueous iodine will occur about 50 percent as I_2 and 50 percent as HOI, thus providing a broad spectrum of germicidal capability.

However, iodine seems unlikely to become a municipal water supply disinfectant in any broad sense because of its potential adverse health effects, high cost, and restricted availability when compared to chlorine. In addition, EPA does not sanction its use in community water supplies because of concern regarding potential toxic effects from long-term exposure.

Noncommunity Use--Iodine dispensers have been used most commonly for small, noncommunity systems serving transient populations. This type of use will probably continue in the future. In practice, a tank containing iodine crystals is attached to the main water supply line. Valves on the main line and tubing leading to the iodine dispenser are set to divert a small flow through the dispenser, where it becomes saturated with iodine and returns to the main line to mix with the rest of the flow to give the desired concentration of iodine in the supply.

Tests of an iodine dispenser by the U.S. Forest Service indicated that "...iodine is an effective disinfectant and that an effective level of iodine can be provided, with very little dispenser maintenance necessary, in all types of water supplies, including spring-fed and hand-pump supplies."[62]

Bromine

Bromine is very similar to chlorine in its water chemistry--hydrolyzing to HOBr, ionizing to OBr^-, and reacting with ammonia to form bromamines. The major differences are that the same degrees of ionization for HOBr occur at pH values about 1 pH unit greater than for HOCl, and that bromamines appear to be nearly as active germicidally as HOBr. At the same time the bactericidal and virucidal effectiveness of HOBr has been found to be comparable to that of HOCl. As a result, solutions of free aqueous bromine exhibit a sharp decline in germicidal effectiveness with increasing pH only at pH greater than 8.5, whereas the effectiveness of free aqueous chlorine falls off sharply at pH greater than 7.5. In addition, the bactericidal and virucidal potency of bromine is reported to change in only a minor way in the presence of nitrogen in the form of ammonia, whereas aqueous bromine does undergo a rapid breakpoint process with ammonia, so that rapid bromine demand equal to about 20 times the ammonia-N may occur.

In spite of its advantages, it seems unlikely that bromine will become a general substitute for chlorine in municipal water disinfection. Its much greater cost and scarcity are strong drawbacks. There have been no thorough investigations on a plant scale that would bring out engineering problems or indicate the general acceptability of waters treated with bromine. The fact that bromine or bromide is physiologically active to man in some measure may also be a bar to general acceptance. Its effectiveness is known to be poor in the presence of organic materials, and the persistence of a residual in distribution systems is not as good as chlorine or iodine.[63] Finally, some tenuous evidence suggests that the total THM yield would be greater for bromine than for chlorine.[64]

Ultraviolet Radiation

The radiation energy of ultraviolet (UV) rays can be used to kill microorganisms. In order to be effective, however, the electromagnetic waves of UV must actually strike an organism, so that it absorbs energy. There is evidence to indicate that UV radiation, when applied under optimum conditions to a water which meets the chemical and physical standards, but not the bacteriological standards, of the

NIPDWR before the ultraviolet treatment, will produce a safe potable water, but with some risk that cysts would still be present after the treatment. This method of disinfection involves exposure of a film of water, up to about 120 mm thick, to one or more quartz mercury-vapor arc lamps emitting germicidal UV radiation at a wavelength of 253.7 nm. Under appropriate conditions, the ultraviolet devices are potentially effective and reliable and do not introduce any taste or chemicals that can react with other substances in the water. A reduction in flow rates poses no problem; an overdose presents minimum danger and is considered appropriate as a safety factor. A properly designed and installed ultraviolet unit, given adequate supervision and maintenance, may be satisfactory for providing disinfection to water from a well which meets regulatory requirements for location and construction.

In 1966 a policy statement on the use of UV processes for disinfection of water was issued by the U.S. Public Health Service. This document recommended that UV radiation (253.7-nm wavelength) be applied at a minimum dosage of 16,000 microwatt-seconds per square centimeter ($\mu W \cdot s/cm^2$) at all points throughout the water disinfection chamber. Although this dosage is conservative, the effective residence time in a particular device could be less than the calculated residence time, owing to nonuniformity of flow. To some extent this factor may be offset because the water being treated does not flow entirely along the wall of the chamber and will, therefore, be exposed to a somewhat higher dosage of UV radiation than that calculated from intensity measurements at the periphery.

Commercial UV water sterilizers contain a low-pressure mercury vapor lamp with about 85 percent of its light output at wavelengths near 253.7 nm. Water passing through the cylinder in which the lamp is situated is exposed to the UV light. The germicidal effect of the lamp is the result of the light's action on the nucleic acids of the bacteria and depends on the light intensity and exposure time. The inactivation of microorganisms can therefore be expressed as a product of intensity, I, of the UV energy, and time, T. The $I \cdot T$ product for UV is analogous to the $C \cdot T$ product for other disinfectants.

Giardia lamblia cysts have been found to be resistant to high dosages of germicidal UV radiation.[65,14] Therefore, all UV water treatment devices that treat raw water in which such cysts may be present should be installed with a filter capable of removing particles greater than 5 μm in diameter.

No UV device should be used on raw water so polluted that it requires extraordinary measures for purification. This is true because of the shielding of microorganisms from the UV light by particulate matter or by clustering of bacteria. Therefore, an upper limit of 1,000 total coliforms/100 mL or 100 fecal coliforms/100 mL has been suggested in Canada for treatment of raw water by UV light.

UV light leaves no residual disinfectant. Therefore, regrowth by photoreactivation may occur when water is exposed to light. Since the water in a distribution system beyond the point of disinfection may be subject to bacterial contamination, a residual disinfectant is recommended in drinking water supplies. If UV radiation is used in such cases, it must be followed by halogen treatment to provide the necessary residual.

Because of the numerous safeguards required in an acceptable UV unit and in its installation, the supervision and maintenance necessary for proper operation, the potential of substances common in many waters that interfere with UV disinfection, and the lack of a simple quick test of the finished water for effective treatment, UV radiation probably will not be frequently used for the disinfection of public water supplies.

Performance--

Various investigations have shown a wide range of UV radiation effectiveness and of sensitivity of different microorganisms to UV energy.

Laboratory Research--Kawabata and Harada, in 1959, reported the following contact times required to achieve 99.9 percent kill at a fixed intensity:[66]

E. coli	60 seconds
Shigella	47 seconds
S. Typhosa	49 seconds
Streptococcus faecalis	165 seconds
B. subtilis	240 seconds
B. subtilis spores	369 seconds

In 1965, Huff et al reported more than four logs (orders of magnitude) of inactivation of polio-, echo-, and coxsackie viruses.[67] The intensities varied from 7,000 to 11,000 μW•sec/cm^2.

Until 1949 it was generally believed that when bacterial cells were exposed to UV light the DNA of these cells were permanently destroyed. In that year, Kelner reported partial reactivation of bacterial cells when exposed to visible light after absorbing UV light in what were thought to be lethal doses.[68] Work by other investigators has confirmed Kelner's findings, and has further shown that the damage caused by UV radiation is the dimerization of thymine, a component of DNA.[69,70] In some organisms, this dimerization can be reversed, if it is not too extensive, by exposure of the cells to visible light in the range of 330 to 500 nm.

Rice and Hoff described work by others showing that UV radiation is ineffective, at the capacities of most commercially available UV treatment units, for inactivation of *Giardia*.[65]

UV light has been tested for effectiveness in killing *Legionella pneumophila* in a model plumbing system containing hot (25°C and 43°C) water. The UV apparatus used was a flowthrough, stainless-steel-enclosed unit rated at 30,000 μW•sec/cm^2 operating at a wavelength of 254 nm. The experimental unit efficiently inactivated *L. pneumophila* by more than 4 logs (99.99 percent) within 60 minutes at both temperatures. Inactivation efficiency was also apparently unaffected when turbid water (suspended solids concentration = 4 to 5 mg/L) was used instead of plain tap water.[80] The suspended solids concentrations used are approximately equivalent to turbidities of 2.5 to 3.5 NTU.

The EPA explored the use of UV for small public water supplies in 1975 through a demonstration grant designed to compare ozone and UV with chlorine as alternative disinfectants.[71] The report results were unfavorable toward UV because of (1) higher costs, (2) lower reliability, and (3) lack of a residual disinfectant.

Heat Treatment

The important waterborne diseases are not known to be caused by spore-forming bacteria or other heat-resistant organisms, and therefore water can be disinfected by subjecting it to heat. Emergency "boil water" orders issued by many utilities in times of emergency are predicated upon the fact that water can be rendered potable by subjecting it even briefly to boiling temperature. The method is, however, impractical on a routine or large-scale basis.

Analogous to the long-established procedures for pasteurization of milk and milk products, there is a continuous-flow water pasteurizer applicable for treating farm ponds, cisterns, and other similar individual domestic water-supply sources. The heart of the 250-gal-per-12-hr unit is a heat exchanger that recovers all but 10°F of the temperature required for treatment and that allows pasteurization of water at 161°F for 15 seconds. Demonstrated effective even with a very high bacterial concentration in the raw water, total cost of household-scale treatment is estimated at about $1 per 1,000 gallons. The principal advantages indicated are reliability and simplicity; disadvantages are the lack of residual disinfection action and relatively high cost.

ADDITIONAL TECHNOLOGIES

The following is a list of Additional Technologies:

- Silver
- Gamma Radiation
- Insoluble Ion Exchange Resins

These disinfection technologies and several others have theoretical potential for use in, at least, limited situations. Unfortunately, little is available in terms of well-documented data regarding their effectiveness, and each is known to have one or more significant disadvantages. A summary of the characteristics of several additional disinfection technologies is presented in Table IV-17.

TABLE IV-17. SUMMARY OF ADDITIONAL TECHNOLOGIES FOR DISINFECTION OF DRINKING WATER

Disinfection Agent*	Technological Status	Efficacy in Demand-Free Systems**			Presence of Residual in Distribution System
		Bacteria	Viruses	Protozoan Cysts	
Ferrate	No reports of use in drinking water	++	+++	NDR†	Poor
High pH conditions (pH 12-12.5)	No reports of large-scale use in drinking water	+++	+++	NDR†	Feasibility restricted since consumption of high pH water not recommended
Hydrogen peroxide	No reports of large-scale use in drinking water	±	±	NDR†	Poor
Ionizing radiation	No reports of use in drinking water	++	++	NDR†	No residual possible
Potassium permanganate	Limited use for disinfection	±	NDR†	NDR†	Good, but aesthetically undesirable
Silver††	No reports of large-scale use in drinking water	+	NDR†	+	Good, but possible health effects

*The sequence in which these agents are listed does not constitute a ranking.
**Ratings: ++++, excellent biocidal activity; +++, good biocidal activity; ++, moderate biocidal activity; +, low biocidal activity; ±, of little or questionable value.
†NDR = Either no data reported or only available data were not free from confounding factors, thus rendering them not amenable to comparison with other data.
††MCL 0.05 mg/L because of health effects (Reference 71).
Source: National Academy of Sciences. "Drinking Water and Health, Vol. 2." National Academy Press, Table II-26, 1980.

DISINFECTION WITH FILTRATION

Descriptions and summaries of the performance of the Most Applicable Technologies indicate that they are capable of inactivating both Giardia and viruses. In terms of C•T values, however, chloramines are much less efficient than the other three Most Applicable Technologies for both microorganisms. In addition, C•T values for chlorine, chlorine dioxide, and ozone are significantly smaller, i.e., inactivation is easier, for enteroviruses than for Giardia cysts. In terms of percent inactivation, it appears practical for chlorine, chlorine dioxide, or ozone to achieve greater than 99 percent inactivation of Giardia cysts and more than 99.9 percent inactivation of viruses, assuming appropriate physical conditions (turbidity, pH, and temperature). Conditions when a water system may be able to rely on disinfection without filtration are described at the end of Section II. Estimated costs for meeting disinfection-only requirements of the Safe Drinking Water Act are contained in Appendix C.

The performance of filtration processes, without disinfection, have been described in detail in Section III. Information presented there shows that filtration processes are all very efficient at removing Giardia cysts, because of their large size, and could be relied on to remove 99.9 percent of these cysts under proper operating conditions. In addition, filtration processes, while not as efficient at removing viruses, can also provide 90 to 99.9 percent of those organisms.

If disinfection is added to filtration processes, it can then be ensured that 99.99 percent or more of viruses present, and 99.9 percent or more of Giardia cysts present can be inactivated or removed from source waters, assuming proper design, operation, and source water quality conditions.

V. Small Water Systems

GENERAL

In this document small water systems are defined as those with design capacities less than 1.0 mgd. They may serve either community systems, or noncommunity systems, and often have distinctly different characteristics and problems than larger plants.

Community Systems

Community systems may be as small as 15 connections or 25 residents---served on a year-round basis. The possible types of communities served by these systems include mobile home parks, subdivisions, small unincorporated communities, training centers, and medical rehabilitation centers.

Noncommunity Systems

Noncommunity systems, as defined by the National Interim Primary Drinking Water Regulations, are water systems that serve transient populations. Included in this system category are hotels, motels, restaurants, campgrounds, service stations, and other public facilities with at least 15 service connections or that serve water to a daily average of at least 25 persons. In addition, some schools, factories, and churches are part of the noncommunity system category.

Water Requirements

Water requirements for small systems may vary over a broad range, depending upon the nature and size of the facility or system served. Based upon a commonly accepted residential per capita water use value of 100 gallons per capita per day (gpcd) for strictly household use, a system serving 25 persons would require a daily flow of only 2,500 gallons. Demands for landscape irrigation and other outside uses could increase the per capita flows by a factor of 6 to 10 above those solely for in-house uses. Site-specific requirements greatly impact the water supply needs for a residential community or noncommunity.

As indicated above, small water systems (community or noncommunity) are defined for this document as those with design flow requirements less than 1.0 mgd. Two surveys of small systems were performed in connection with this report to determine characteristics such as system supply capacity, treatment design capacity, average day flow requirements, and system storage capacity.[1,2] The results of those two surveys, coupled with data collected informally from other small water systems, lead to the definition of flow characteristics for four flow categories for the purposes of this report, as defined in Table V-1 below. Flow characteristics of larger systems are defined in Section VI.

TABLE V-1. SMALL SYSTEM SUPPLY CHARACTERISTICS

Category	Average Daily Flow, mgd	Treatment Design Capacity, mgd	Storage Capacity, MG	System Supply Capacity, mgd
1	0.013	0.026	0.044	0.070
2	0.045	0.068	0.082	0.150
3	0.133	0.166	0.174	0.340
4	0.400	0.500	0.340	0.840

As shown in Table V-1, small systems often have large ratios of system supply capacity to average daily flow, e.g. 5.4:1 for Category 1. At the same time, however, they generally satisfy that demand through a combination of treatment capacity and storage capacity. The ratios of treatment design capacity to average flow used for these purposes are as follows:

Category	Ratio of Treatment Capacity to Average Flow
1	2:1
2	1.5:1
3	1.25:1
4	1.25:1

Storage capacities shown in Table V-1 are the difference between system supply capacity values (data from surveys) and treatment design capacities developed from the above ratios.

WATERBORNE DISEASE OUTBREAKS

As described in Section II of this report, noncommunity systems are normally the source of about as many disease outbreaks as community systems every year. In addition, noncommunity systems are significant simply because of the large number of such systems in this country (recently estimated at about 150,000).[3]

The number of disease outbreaks in noncommunity systems also usually exceeds those in community systems during the summer months. As indicated by the definition of noncommunity systems, they serve a large number of recreational areas experiencing heavy use for the summer, causing both maximum demand on water systems and frequent overloads to sewerage systems.

While the reported number of outbreaks in noncommunity systems is high (Section II), it is thought that the problem of under-reporting of outbreaks is worse for noncommunity systems than for community systems. For example, outbreaks in noncommunity systems are difficult to detect because they serve travelers who often become ill far from the point of exposure to contaminated drinking water. Many of these people, therefore, do not associate their illnesses with a place they visted days earlier.

TREATMENT FACILITIES USED BY SMALL WATER SYSTEMS

The processes and facilities used by existing small water systems to treat surface water vary about as widely as the range of flows they treat. As an example of the treatment processes used, a survey of seven small systems in the northeastern United States found that two used disinfection only, one used slow-sand filters and disinfection, one used gravity sand filters (without pretreatment) and disinfection, and three used small conventional treatment systems.[2]

A survey of 80 water systems in Wyoming provides similar results regarding small system treatment processes to those of the seven northeastern plants mentioned above. For 15 systems in Categories 1 and 2, for example, two were providing no surface water treatment, three were providing disinfection only, and 10 were providing some combination of both filtration and disinfection.[4] The types of filtration systems used at the ten plants included package pressure filter systems, package complete treatment systems, and conventional treatment systems. Most disinfection systems used sodium hypochlorite solution feed, although several used chlorine feed with cylinder storage.

DIFFICULTIES SPECIFIC TO SMALL SYSTEMS

Analysis of the causes of known outbreaks in noncommunity systems indicates that three-fourths of them are related to either use of untreated groundwater or poor treatment practices. Specifically, small community and noncommunity systems commonly suffer from design, construction, operation, and maintenance problems. They also normally receive less attention from regulatory agencies than larger systems, in terms of monitoring and inspection to correct problems.[3]

A symposium on small water systems, held in 1978, cited many of the same problems and causes described above.[5] Several speakers also added related difficulties for small systems, including: (1) most small systems lack funds for proper design and construction; (2) many operators of small water systems have multiple duties; and (3) many such operators are too far away from educational centers to attend courses for licensing.

For campgrounds and similar areas, additional difficulties are usually present. Water supply and treatment sites may be remote, and electricity may not be available.

Small potable water treatment systems must perform satisfactorily under adverse conditions with a minimum of attention.

TREATMENT TECHNOLOGIES APPLICABLE TO SMALL SYSTEMS

Many of the technologies described in Section III, Filtration, and Section IV, Disinfection, are adaptable to smaller systems. Others, because of such factors as operational complexity, safety considerations, equipment size limitations and cost, are not appropriate for small systems.

Filtration technologies that can serve small systems include:

- Package plants
- Slow-sand filters
- Diatomaceous earth filters
- Ultrafiltration
- Cartridge filters

Disinfection technologies which may be appropriate for small systems include:

- Hypochlorination and gaseous chlorination
- Iodination
- Erosion feed chlorinators
- Ultraviolet radiation
- Ozonation

The treatment and performance capabilities of the above filtration and disinfection technologies are described in detail in Sections III and IV of this document. Following are additional discussions of the specific technologies as they relate to small systems.

FILTRATION TECHNOLOGIES

Package Plants

When a surface water is the only source of supply, it must be treated with equipment that can accommodate the wide range of raw water quality that may occur, and usually, with a minimal amount of operator attention.

Many parks located at lakes and reservoirs have used complete-treatment package plants such as those described in Section III with excellent success.

A survey of 17 package water treatment plants, ranging in capacity from 14,000 gpd to 80,000 gpd, was conducted from 1977 through 1979.[6] These plants served facilities varying from a private ski resort to several U.S. Army Corps of Engineers reservoir park sites. A national survey of 27 package plants was conducted in 1986.[7]

The application requirements for adapting these package plants to small systems are similar to those for large systems. Of major concern is the quality and amount of operational attention these plants will receive. To minimize required operator skill level and operational attention, equipment should be automated. Continous effluent turbidity and chlorine residual monitoring systems with alarms and emergency shutdown provisions are features that safeguard water quality and should be provided for unattended plants.

Slow-Sand Filters

Slow-sand filters are applicable to small water supply systems and should be extremely attractive where no power is available. Their proven record of effective removal of turbidity and *Giardia* cysts makes them suitable for application where operational attention is minimal. Since no chemicals other than chlorine are needed, and they involve no mechanical equipment, the required operator skill level is the lowest of the filtration alternatives available to small systems. Section III contains an extensive discussion on performance of slow-sand filters.

Diatomaceous Earth Filters

Diatomaceous earth (DE) pressure and vacuum filters can be used on relatively low turbidity surface waters (less than 1 to 2 NTU) for removal of turbidity and Giardia cysts. DE filters can effectively remove particles as small as 1 micron, but would require coagulating chemicals and special filter aids to provide significant virus removal.

A small DE filter has been in service for more than 6 years at Van Duzer Forest Corridor State Wayside in Lincoln County near the Pacific Ocean in western Oregon.[8] The system processes springwater collected in a 20,000-gallon tank, from which it flows by gravity to the DE filter installation. A small pressurizing pump feeds water to the filter. Treated water is stored in a 400-gallon tank for distribution to park facilities. Filtered effluent is continously monitored for turbidity, and when a value of 1 NTU is reached the media is changed.

Ultrafiltration

Ultrafiltration is a process that utilizes a hollow fiber membrane to remove undissolved, suspended or emulsified solids from water supplies without requiring coagulation. It is most generally used in specialized applications requiring extremely high purity water, as pretreatment prior to reverse osmosis or for removal of colloidal silica from boiler feed water. Application to potable water treatment is limited to extremely high quality raw water supplies of low turbidity (1 NTU or less), or following pretreatment to produce a supply of low turbidity.

The hollow fiber membrane, which operates over a pressure range of 10 to 100 psig, excludes particles larger than 0.01 μm. Bacteria, Giardia cysts and virus particles are effectively removed by the fiber membranes. The hollow fiber membranes are contained in a pressure vessel or cartridge. During operation, the flow is introduced to the inside of the fiber. Filtrate or permeate collects on the outside of the membrane. Concentrate exits at the end of the hollow fiber and is discharged to waste. In a typical single-pass application, 90 percent of the feed water is recovered as permeate. The membranes are cleaned by backflushing, which restores the original porosity and allows continuous use for indefinite periods. In some applications a chlorine solution is added to the backflushing flow.

Ultrafiltration membranes are sensitive to the concentration of suspended and colloidal solids in the feed water. The flux level (gallons of permeate produced per square foot of membrane, GSFD) and the flux stability are influenced by the feed water quality, the filtering cycle duration, and the quality of water used for backflushing the membranes. Where a water supply has a Fouling Index of 10 or less (typical of most groundwaters and high clarity surface water), filtering cycles will range up to 8 hours with about a 10 percent reduction in flux. In general, capacity can be restored to nearly clean membrane values by a fast-forward flush of influent to waste. However, after two or three flushings, the membranes must be backwashed to restore the initial flux level.

A typical ultrafiltration installation for a small water supply system would include a skid-mounted ultrafiltration unit containing the hollow-fiber membranes in cartridges, automatic and manual valves for backwashing and unit isolation, flow meters, pressure gauges, integral backwash pump, and control panel. A separate supply pump would be required, as would all interconnecting piping serving plants with multiple units. Storage tanks and chemical feed pumps for membrane cleaning solutions would be needed. Product water storage with chlorination provisions would also be required.

Cartridge Filters

Cartridge filters using microporous ceramic filter elements with pore sizes as small as 0.2 μm may be suitable for producing potable water from raw water supplies containing moderate levels of turbidity, algae and microbiological contaminants.[9] Single-filter elements may be manifolded in a pressurized housing to produce flow capacities approaching 24 gpm (34,000 gpd) from a single assembly. The clean filter element pressure drop is about 45 psi at maximum capacity. The ceramic elements are cleaned when the pressure drop reaches 88 psi.

Cleaning is accomplished by opening the filter housing and scrubbing each individual vertical filter element with a hand-operated hydraulically driven brush that fits over each element. An additional feature offered with one manufacturer's equipment is the incorporation of finely divided particles of silver within the ceramic matrix, which prevents the growth of bacteria in the element. Other manufacturers utilize disposable polypropylene filter elements in multi-cartridge stainless steel housings. These units are available in capacities ranging from 2 gpm to 720 gpm. Membrane pore sizes range from 0.2 μm to 1.0 μm in equipment suitable for producing potable water. A disadvantage of the polypropylene membrane is that it may be cleaned only once and then must be replaced. However, according to manufacturers' guidelines, filtration of a low turbidity surface water (2 NTU or less) could be expected to provide service periods between filter element replacement ranging from 5 to 20 days, depending upon the pore size of the cartridge selected.

The application of cartridge filters using either cleanable ceramic or disposable polypropylene cartridges to small water systems appears to be a feasible method for removing turbidity and most microbiological contaminants, although data are needed regarding the ability of cartridge filters to remove viruses. The efficiency and economics of the process must be closely evaluated for each application. Pretreatment in the form of roughing filters (rapid sand or multi-media) or fine mesh screens may be needed to remove larger suspended solids, which could quickly foul the cartridges, reducing capacity. Prechlorination is recommended to prevent microbial growth on the cartridges and to inactivate any organism that might pass through the filter elements, including viruses.

The advantage to small systems, is that, with the exception of chlorination, no other chemicals are required. The process is one of strictly physical removal of small particles by straining as the water passes through the porous membranes. Other than occasional cleaning or membrane replacement, operational requirements are not complex and do not require skilled personnel. Such a system would be suited ideally to many small systems where, generally, only maintenance personnel are available for operating water supply facilities.

Long analyzed a variety of cartridge filters for removal efficiency by using turbidity measurements, particle size analysis, and scanning electron microscope analysis.[10] The filters were challenged with a solution of microspheres averaging 5.7 μm in diameter (smaller than a Giardia cyst), at a concentration of 40,000 to 65,000 spheres per mL. Ten of 17 cartridge filters removed over 99.99 percent of the microspheres, two others removed 99 to 99.9 percent, and five were not evaluated for microsphere removal because they did not have turbidity reductions greater than 90 percent.

In tests using live infectious cysts from a human source, cartridge filters were found to be highly efficient in removing Giardia cysts.[11] Each test involved challenging a filter with 300,000 cysts. The average removal for five tests was 99.86 percent, with removal efficiencies ranging from 99.5 percent to 99.99 percent.

DISINFECTION TECHNOLOGIES

The various disinfection technologies are described in detail in Section IV. Several of the technologies classified as Other Applicable Technologies and Additional Technologies are appropriate for small systems. These are discussed in the following paragraphs as they relate to small systems.

Hypochlorination and Gaseous Chlorination

Chlorination is the most widely used and accepted technique for disinfection in small and large systems alike. In small systems, both gas chlorinators and hypochlorite solution feeders are used. Gas chlorination is discussed in Section IV of this report. Chlorine in the hypochlorite form is often used because it is easy and safe to handle for unskilled personnel, and the application equipment is generally inexpensive. Hypochlorite solution made from either calcium or sodium hypochlorite is commonly used.

For small systems, hypochlorination facilities generally consist of an electrically operated, diaphragm-style chemical metering pump, a solution storage/mixing tank, and the necessary controls. Solutions of hypochlorite are fed to the system at a constant rate, controlled by level or flow switches. In more complex systems, dosages may be paced directly to flow using a water meter to proportion solution feed to flow. Typically, the former means of control is used in small systems. At constant feed rate, if flows change, chlorine dosage levels also change, leading to the possibility of inadequate disinfection.

In remote facilities where power is unavailable, a positive displacement style feed pump powered by a specially designed water meter, which proportions hypochlorite solutions directly to flow, may be used. Such a system would be appropriate where water is collected at a spring and then conveyed by gravity to the point of use. A minimum water pressure of 15 psi is needed to properly operate this type of feeder.

Iodination

Iodine dispensers have been used most commonly for small noncommunity systems and have been evaluated extensively by the U.S. Forest Service.[12] A detailed discussion of the equipment, its application and use can be found in Section IV.

Erosion Feed Chlorinators

Erosion feed chlorinators are simple nonmechanical devices that are suitable for small systems. They operate by erosion of calcium hypochlorite tablets at a controlled rate. The chlorinator consists of a refillable hopper, which stores the hypochlorite tablets and immerses them into a dissolving chamber. The water level in the dissolving chamber establishes the erosion rate; hence the hypochlorite solution strength. The solution is then metered into a contact tank or reservoir by gravity, with flow through the chlorinator controlled by water withdrawn from the contact tank.

These systems are mechanically simple, but require frequent adjustment and close control to maintain continuous and uniform hypochlorite dosages in the water supply. They are advantageous where no power source is available.

Testing was conducted by the U.S. Forest Service on one erosion chlorinator model.[13] The chlorinator was shown to provide unstable dose rates when operated in the continuous flow mode. Intermittent-flow operation provided a more stable dose rate, and that mode of operation was recommended as the only one to be used for erosion chlorinators.

Ultraviolet Radiation

The application of ultraviolet (UV) radiation for disinfection of water supplies is discussed in Section IV. The following discussion relates to the use of this disinfection technology in small noncommunity systems.

It is pointed out in Section IV that UV radiation has limited general use in the disinfection of public water supplies principally because the lack of a residual presents the opportunity for microbial contamination within the distribution system. This may not be an insurmountable problem in small noncommunity systems where distribution system piping is minimal.

The application of UV radiation for disinfection must be approached with caution. The effect of UV radiation is different for various microorganisms. Of particular importance are Giardia cysts, which require higher dosages than can be provided by some UV units. Where Giardia cysts are a potential problem, use of UV for disinfection should be carefully evaluated.

Unfortunately, there are no uniform standards for UV light disinfection. Equipment manufacturers customarily rate their equipment in terms of dose and flow rate. Dose is usually reported as energy per unit area in microwatt-seconds per square centimeter ($\mu W \cdot s/cm^2$). Between different equipment models and manufacturers, dose can range from as low as 16,000 to as high as 70,000 $\mu W \cdot s/cm^2$. Along with this wide range in dose, different equipment will provide widely varying detention time and water layer thickness. Thus, care must be exercised when selecting UV light disinfection equipment.

Transmission of UV energy through water is essential for adequate disinfection. A low-turbidity water is essential to the use of UV disinfection, and decreasing turbidity is a principal factor in increasing the clarity to UV light. Also, many substances can interfere with UV light transmissivity.

Prior to consideration of this disinfection technology, water samples should be tested to determine whether UV light disinfection is applicable, what additional processes, if any, would be required to use UV light disinfection, and design criteria for the UV equipment.

Ozonation

Small-scale ozone generation, feed, and contact systems are available in package form for use in small water treatment plants. The technology and equipment for these systems is essentially the same as that described in Section IV.

ALTERNATIVES TO TREATMENT

Under certain circumstances, some small systems may have alternatives available to them that are not practical for larger systems. Specifically, it may be possible for a small system to construct a well to provide a groundwater source as either a supplement to or a replacement for an existing surface water source. Further, some systems may be small enough that purified water vending machines could be used to supply all or a major portion of its water demand.

Wells

Depending on their capacity and depth; a well can be either a very economical or very expensive alternative to surface water treatment. Once constructed, the well would provide a groundwater source which, in most circumstances, would require less treatment than a surface water supply. As a safeguard, construction of any new well should be preceded by chemical and bacteriological testing in addition to hydraulic testing. Following construction of the well, operation and maintenance will be required on the pumps and motors at the well site, and disinfection facilities should also be provided. Costs for groundwater disinfection are included in Appendix A of this report.

Purified Water Vending Machines

An additional alternative for surface water treatment in some small water systems can already be found at many supermarkets--vending machines that provide purified water in bottles. Machines of this kind are included here because they could be purchased and maintained by a water system. A bottled-water delivery service, on the other hand, could not be controlled by the water system management and is therefore not included in this definition.

Purified water vending machines contain treatment processes that can remove microbial contaminants when connected to an existing water supply. The reliability of the treatment processes would depend on both adequate maintenance procedures and schedules, and on internal and external monitoring of the quality of water produced by the machines. A number of vendors now produce these machines, which often include ion exchange and reverse osmosis in addition to mechanical filtration, granular activated carbon, and ultraviolet light for disinfection.

The efficiency and reliability of these machines must be checked for each vendor, and vigilant maintenance and monitoring is required.

VI. Cost Data

BASIS OF COSTS - GENERAL

Capital and operating costs for the technologies in this document are based upon updated costs originally presented in several cost documents prepared for EPA.[1,2] Construction cost information originating from those reports was modified and updated by acquisition of recent cost data. General cost information is presented in the paragraphs immediately following. Specific details regarding cost calculations for individual processes and process groups are presented later in this section under the heading "Typical Treatment Costs for Surface Water," and in Appendix B. Description of assumptions and costs for disinfection of groundwater are presented in Appendix A.

Capital Costs

The construction cost data presented in this report are based on late 1986 costs. Methods of calculating construction costs are described in the two EPA reports mentioned above, References 1 and 2. In general, the construction cost for each unit process is presented as a function of the most applicable design parameter for the process. For example, construction costs for gravity and pressure filters are presented versus total square feet of filter surface area; ozone generation system costs are presented versus pounds per day of feed capacity. Use of such key design parameters allows the cost data to be utilized with greater flexibility than if cost information were plotted versus flow. However, for some processes, flow is the most appropriate design parameter. An example of such processes is ultraviolet light disinfection.

Construction costs are converted to capital costs by use of typical percentages of construction costs for such factors as contingencies (15%), contractor's overhead and profit (12%), sitework (15%), subsurface considerations (5%), engineering and technical fees (15%), and interest during construction (10%). Costs for acquiring new land for treatment sites, however, are not included in project costs since these costs are extremely site and project specific, since: (1) many water utilities already own land which could satisfy all or a portion of their future needs, and (2) the unit cost of land is highly variable from one location to another. To estimate annual costs and costs per 1,000 gallons, capital costs were amortized over 20 years at a 10 percent interest rate.

Operation and Maintenance Costs

Operation and maintenance requirements in the computer model were developed for energy, maintenance material, and labor. The energy category includes process electrical energy, building electrical energy, and diesel fuel. The operation and maintenance requirements were determined from operating data at existing plants, to the extent possible. Where such information was not available, assumptions were made based on the experience of both the author and the equipment manufacturer, and such assumptions are stated in the individual process write-ups.

Electrical energy requirements were developed for both process energy and building-related energy, and they are presented in terms of kilowatt-hours (kWh) per year. Process energy is energy required for operation of motor controls and instrumentation. Building energy is energy required for lighting, ventilation, and heating of buildings that house treatment processes.

Maintenance material costs include the cost of periodic replacement of component parts necessary to keep the process operable and functioning. Examples of maintenance material items are valves, motors, instrumentation and other process items of similar nature. Maintenance material requirements do not include the cost of chemicals required for process operation. Chemical costs are added separately, as shown in the project costs at the end of this section. Unit costs of chemicals used in this report represent late 1986 conditions, and are shown in Table VI-1. Labor requirements include both operation and maintenance labor, and are presented in terms of hours per year in the cost documents serving as the basis for this work.[1,2]

TABLE VI-1. CHEMICAL COSTS USED

Chemical	Small (<1 mgd) Systems, $/Ton	Large (>1 mgd) Systems, $/Ton
Alum (Dry)	$ 500	$ 250
Alum (Liquid)	300	125
Lime (Quick)	100	75
Lime (Hydrated)	150	100
Ferric Chloride	500	275
Ferrous Sulfate	277	250
Ferric Sulfate	200	155
Soda Ash	250	200
Sodium Hydroxide	595	316
Chlorine	500	300
Sodium Hypochlorite	190	150
Liquid Carbon Dioxide	350	100
Sodium Hexametaphosphate	1,160	1,100
Zinc Orthophosphate	1,520	1,000
Ammonia, Aqua	230	200
Ammonia, Anhydrous	410	350
Sulfuric Acid	140	100
Hydrochloric Acid	171	166
Activated Carbon Powdered	950	800
Activated Carbon granular	1,900	1,600
Activated Alumina	1,694	1,156
Potassium Permanganate	2,800	2,500
Sodium Bisulfate (Anhyd)	909	.673
Sodium Silicate	400	200
Sodium Chloride	105	85
Polyelectrolyte	1,500	1,000
Diatomaceous Earth	680	310
Magnesium	650	582
Sodium Chlorite	3,200	2,800
Sodium Hydroxide 76%	590	316
Sodium Bicarbonate	490	380
Calcium Hypochlorite	2,700	1,540

Total operation and maintenance cost is a composite of the energy, maintenance material, and labor costs. To determine annual energy costs, unit costs of $0.086/kWh of electricity and $0.80/gallon of diesel fuel are used. The labor requirements are converted to an annual cost using hourly labor rates of $5.90/hour for small water systems (less than 150,000 gpd design capacity) and $14.30/hour for larger systems, based on labor costs found at existing plants during conduct of this project.

Updating Costs to the Time of Construction

Continued usefulness of the cost data in this report requires updating to reflect increases or decreases in the prices of the various components.

The simplest, and most universally used method of updating construction costs is through use of the ENR (Engineering News Record) Construction Cost Index (CCI). To update construction costs in this report by using the CCI (1967 average = 100), which was assumed to be 405 in late 1986, the following formula may be utilized:

$$\text{Updated Cost} = \text{Construction Cost}\ \left(\frac{\text{Current CCI}}{405}\right)$$

The computer model used in this report uses Bureau of Labor Statistics (BLS) indices for a number of construction cost components, including manufactured equipment, concrete, steel, piping, valves, and electrical instrumentation. ENR indices are used for several other construction cost component categories. BLS indices may be found in Table 6 of the publication, "Producer Prices and Price Indexes," while ENR indices may be found in issues of **Engineering News Record** magazine.

Updating of operation and maintenance costs may be accomplished by updating the three individual components: energy, labor, and maintenance material. Energy and labor are updated by applying the current unit costs to the kilowatt-hour, diesel fuel, and labor requirements obtained from the energy and labor curves. Maintenance material costs, which are presented in terms of dollars per year, can be updated using the Producer Price Index for Finished Goods.

BASIS OF COSTS - PROCESS BY PROCESS

The following pages contain descriptions of the processes used in treatment technologies used to remove microbial contaminants from drinking water. The description of each process, or treatment system component, includes a discussion of both the conceptual design of structures and equipment, and the operation and maintenance requirements of the process or component. Chemical use requirements, where appropriate, are described separately.

Process and component descriptions are divided into five groups and 12 size categories on the following pages. The five process groups are:

- Pumping
- Chemical feed
- Filtration process components
- Disinfection processes
- Solids handling processes

The 12 size categories apply to treatment systems rather than individual processes and represent ranges of populations that could be served by a treatment plant of a given size. The categories, population ranges, plant capacities, and average flows shown in Table VI-2 were selected after a national survey of operating treatment plants.[3] Capacities range from 26,000 gpd for the smallest plant to 1.3 billion gpd for the largest. Construction costs are based on plant capacity flow rates shown in the table, while operation, maintenance, and chemical use costs are based on average flow rates.

Pumping

Eight types of pumping systems may be used with the treatment processes described in this section. The eight types are:

- Package Raw Water Pumping
- Raw Water Pumping

TABLE VI-2. PROPOSED AVERAGE PRODUCTION RATE AND PLANT CAPACITIES

Category	Population Range	Average Flow (Q_A), mgd	Surface Water Treatment: Plant Capacity, mgd	Surface Water Treatment: Q_A as % of Plant Capacity
1	25 - 100	0.013	0.026*	50.0
2	101 - 500	0.045	0.068*	66.2
3	501 - 1,000	0.133	0.166*	80.0
4	1,001 - 3,300	0.40	0.500*	80.0
5	3,301 - 10,000	1.30	2.50	52.0
6	10,001 - 25,000	3.25	5.85	55.6
7	25,001 - 50,000	6.75	11.58	58.3
8	50,001 - 75,000	11.50	22.86	50.3
9	75,001 - 100,000	20.00	39.68	50.4
10	100,001 - 500,000	55.50	109.90	50.5
11	500,001 - 1,000,000	205	404	50.7
12	>1,000,000	650	1,275	51.0

* Costs for supply systems in Categories 1-4 include significant supplemental storage volumes. See Section V.

- In-Plant Pumping
- Backwash Pumping
- Package High-Service Pumping
- Finished Water Pumping
- Unthickened Chemical Sludge Pumping
- Thickened Chemical Sludge Pumping

Package Raw Water Pumping --

Conceptual Design--Construction cost estimates are for raw water pumping facilities with capacities in Categories 1, 2, and 3. Costs are based on the use of premanufactured package stations using duplex submersible pumps contained in a 20-foot-deep steel pump sump. The sump is covered, and includes an entry hatch and ladder rungs for operator access. The sump is supported on a concrete anchor slab at the bottom of the excavation. The pumping facilities are located adjacent to a stream or lake, and water enters the pumping station by gravity flow from the intake structure. The facilities are capable of pumping against a head of 50 feet.

The complete system includes the pump sump, the premanufactured package pumping station, manifold piping within the sump, sump intake line valve, pump check valves, and electrical controls. Excavation costs are based upon a 1:1 side slope, and allowance for dewatering and backfill compaction. Main electrical control panels are included in the cost estimate, and these panels are assumed to be located in the treatment plant control building. Excluded are costs of the raw water intake structure, and transmission lines between water source and pump sump and between the pump sump and the treatment facilities. Costs of these items are excluded because specific site conditions will result in significant variations in requirements and cost. No housing for the pumping facilities is required.

Operation and Maintenance Requirements--Process electrical energy requirements are for continuous 24 hour/day operation of raw water pumps at a TDH (total dynamic head) of 50 feet. Power requirements are based on a pumping efficiency of 80 percent and a motor efficiency of 90 percent. Since the facilities are not in a building, no energy for heating or ventilating is required. Lighting requirements in the sump are minimal and are not included.

Annual maintenance material requirements for submersible pumps, valves, and electrical equipment are estimated at 1 percent of the construction costs of these items. Labor requirements are for general maintenance of pumping station equipment. Maintenance requirements for the totally sealed submersible pumps are minimal.

Raw Water Pumping--

The costs for raw water pumping stations for Categories 4 through 12 are based on different assumptions than for package pumping stations, as described below.

Conceptual Design--Raw water pumping facilities exclude a wetwell and housing, since these requirements will be extremely variable from location to location. The facilities include pumps, valves, manifold piping, and electrical equipment and instrumentation. Pumps are constant-speed vertical turbine type, driven by dripproof, high-thrust vertical motors, assumed to pump against a TDH of 100 feet. A standby pump equal in capacity to the largest pump is also included. Manifold piping is sized for a velocity of 5 ft/sec.

Operation and Maintenance Requirements--Process energy requirements are calculated for a 100-foot TDH, using a motor efficiency of 90 percent and a pump efficiency of 85 percent.

Maintenance materials include repair parts for pumps, motors, valves, and electrical motor starters and controls. Labor requirements are based on operation and maintenance of the pumps, motors, and valving, plus maintenance of electrical controls.

In-Plant Pumping--

In-plant pumping is required in most water treatment plants, except for those using pressure filtration, and for package treatment plants that contain their own internal pumping.

Conceptual Design--Pumps for this purpose are capable of pumping against a TDH up to 75 feet. The pumps are constant speed, vertical turbine pumps driven by drip-proof, high-thrust, vertical motors. A standby pump and motor is included in each installation, together with a wet well, and piping and valves to connect the pumps to a common manifold. Piping was sized for a velocity of 5 ft/sec.

Operation and Maintenance Requirements--Process energy requirements were calculated using a motor efficiency of 90 percent and a pump efficiency of 85 percent. Maintenance material includes repair parts for pumps, motors, valves, and electrical starters and controls. Labor requirements are based on operation and maintenance of the pumps, motors, and valving, plus maintenance of electrical controls.

Backwash Pumping--

For purposes of developing filtration treatment costs in this report, it is assumed that backwash pumping, as a separate plant component, will be used where system flow is Category 3 and larger. Backwashing for Categories 1 and 2 is assumed to be included in package filtration facilities for treatment systems of that size.

Conceptual Design--The cost of a backwash pumping system must be added to the cost of filtration structure, filter media, surface wash systems, and backwash water storage facilities to develop a complete filtration facility cost. The backwash pumping system cost includes the cost of required pumps and motors, including one standby unit, flow control, filter backwash sequencing control, pumping station valving, the backwash header cost not included in the filter structure, and motor starters. Backwash piping and valving is sized for a velocity of 7 ft/sec. Housing costs are not included. The assumed pumping head for the backwash pumps is 50 feet (TDH), and the maximum design rate for backwash is 18 gpm/ft^2. The largest pump utilized is 7,000 gpm, and one standby pump is included for all installations.

Operation and Maintenance Requirements--A backwash frequency of two per day with a 10-minute duration per wash is assumed. For dual-cell filters, a backwash is defined as a backwash of both cells. Filters are assumed to be sized to receive 155 percent plant design flow at 5 gpm/ft^2.

Energy requirements are calculated using a backwash rate of 18 gpm/ft^2, a pumping head of 50 feet (TDH), and an overall motor/pump efficiency of 70 percent. Energy requirements are based on a backwash period of 10 minutes twice a day.

Maintenance material costs are for repair of the backwash pumps, motor starters, and valving. Labor requirements are for maintenance labor only, as all operation labor is included in labor related to the filtration structure.

Package High-Service Pumping--
Package high-service water pumping facilities are used to supply finished water at a relatively uniform pressure to a system with fluctuating demand. These facilities are appropriate for Categories 1 through 4.

Conceptual Design--Costs are developed for pumping stations that contain pumps, pressure sensing, and flow control valves, and required electrical equipment and instrumentation. Pumping stations utilize two end-suction centrifugal pumps for capacities smaller than 600,000 gpd, and three pumps for greater flows. The pumps provide a maximum output pressure of 70 psi and are designed for flooded suction application.

No allowances for housing costs are included since spatial requirements are minimal. Water storage is supplied by the clearwell tank.

Operation and Maintenance Requirements--Pumping units are selected to handle peak hourly flows and utilize a two- or three-pump system with a lead pump and one or two main pumps. In all systems, continuous 24-hour operation of the lead pump and 8-hour operation of the main pump are assumed to determine energy usage. Pumping costs are computed based on supplying the indicated flows at 70 psi discharge pressure.

Maintenance material costs are related to replacement costs for seals and other miscellaneous small parts. Labor requirements consist of pump seal lubrication, occasional seal replacement, and calibration of pressure control devices.

Finished Water Pumping--

Finished water pumping facilities are assumed to be custom-designed and contractor-constructed facilities for treatment systems in Categories 5 through 12. Smaller treatment facilities are assumed to be served by package high-service pumping systems described above.

Conceptual Design--Depending on distribution system layout and storage, finished water pumping requirements may be greater than the average daily plant flow. To account for such variations in capacity of the finished water pumping facilities, costs are based on facilities with a capacity equal to 150 percent of plant design flow. Pumps utilized are the vertical turbine type driven by 1,800-rpm, constant speed, dripproof, high-thrust vertical motors. One standby pump with capacity equal to the largest pump is also provided.

The facilities include all electrical equipment and instrumentation as well as valving and manifolding within the pumping station. Manifold piping is sized for a velocity of 5 ft/sec. No costs are included for housing or a wetwell, as it was assumed that the clearwell would serve as the wetwell. Separate costs are provided for clearwell storage.

Operation and Maintenance Requirements--Process energy requirements are based on a 300-foot TDH, using a motor efficiency of 90 percent and a pump efficiency of 85 percent. Maintenance materials include repair parts for pumps, motors, valves, and electrical starters and controls. Labor requirements are based on operation and maintenance of the pumps, motors, and valving, plus maintenance of electrical controls.

Unthickened Chemical Sludge Pumping--

Chemical sludge originating in clarifiers or wash water storage basins is assumed to be pumped to dewatering facilities such as thickeners or sludge lagoons.

Conceputal Design--Unthickened chemical sludge pumping stations have separate wet and dry wells. Variable speed, horizontal, centrifugal pumps are used for pumping dilute sludge, and one standby pump is included for each pumping station.

Pipe and valves are sized for velocities of approximately 5 ft/sec. Costs are based on a 12 foot depth and stairway access for stations with capacities greater than 100 gpm. Housing for electrical equipment and access to the dry well is located above the dry well, but the housing does not cover the entire dry well.

Operation and Maintenance Requirements--Process electrical energy costs are based on pumping a lime sludge against a TDH of 30 feet and an overall motor-pump efficiency of 65 percent. Maintenance material requirements are for periodic repair of the pumps, motors, and electrical control units. Labor requirements are for periodic checking of pumps and motors, and for periodic maintenance.

Thickened Chemical Sludge Pumping--
Pumps used to convey sludges from thickening devices to dewatering units have different design concepts and operation and maintenance requirements than pumps for unthickened chemical sludge.

Conceptual Design--Pumping facilities include progressive cavity pumps with the pumps drawing thickened sludge directly from a gravity thickener. The pumps are assumed to be located in a building used for other purposes, such as sludge dewatering or lime recalcination. The estimated costs include the pump and motors, required pipe and valving within the building, electrical equipment and instrumentation, and housing.

Operation and Maintenance Requirements--Process electrical energy costs are based on pumping a thickened lime mud against a TDH of 50 feet and manufacturers' estimates of connected horsepower. Maintenance material requirements are for periodic repair of the pumps, motors, and electrical control units. Labor requirements are for periodic checking of the pumping equipment, and for periodic maintenance.

Chemical Feed

Chemical feed facilities are used as supplements to other filtration-related or disinfection-related processes. The types and quantities of chemicals added to a water supply during treatment depend on the process being used and the quality of the raw water.

Six types of major chemical feed facilities are used with treatment facilities for which costs are provided at the end of this section. The six types are:

- Basic Chemical Feed
- Liquid Alum Feed
- Polymer Feed
- Sodium Hydroxide Feed
- Lime Feed
- Sulfuric Acid Feed

Where other chemicals are required in relatively small quantities for some treatment processes, the cost of the feed facilities for those chemicals are included in the cost of the treatment process.

Basic Chemical Feed--

Feed systems for chemicals such as soda ash, hydrated lime, liquid or dry alum, ferrous sulfate, and sodium hexametaphosphate are very similar. A basic chemical feed system described here can be used to feed these chemicals at water plants for Categories 1 through 4.

Feed systems for disinfectants, sodium hydroxide, and hydrated lime (for Categories 7 through 12), and alum for Categories 5 through 12, are specific to each of these chemicals. Descriptions of systems for feeding these chemicals are presented in subsequent sections.

Conceptual Design--The most common type of chemical feed system consists of a mixing tank, mixer, and metering pump. Chemical solutions of known concentrations are prepared on a batch basis by manually adding a chemical and water to the dissolving tank. This system can be used to feed either liquid or dry chemicals. For dry chemicals, the mixing tank is used for dissolving and mixing the chemicals with water and for storage of the chemical solution. For liquid chemicals, the mixing tank is used for initial mixing and solution storage. A metering pump is used to accurately feed a solution of known concentration at a set rate.

Two basic assumptions made in sizing the storage tank are that it should only be necessary to add chemicals once daily (assuming up to 24 hours/day of operation) and that the chemical should be added to produce a 6 percent solution concentration. Mixer horsepower requirements are based on the manufacturer's recommendations for the various tank sizes. Metering pump capacities are based on the daily volume of 6 percent solution, which correlates with the storage tank size and upon standard sizes provided by manufacturers of package water treatment plants.

Operation and Maintenance Requirements--Operation labor requirements for the basic chemical feed system include labor for emptying bags of dry chemicals, or containers of liquid chemicals, into the dissolving tank for turning the mixer on and off, for calibrating the chemical metering pump, and for occasional preventive maintenance of the mixer and metering pump.

Electrical requirements are for the mixer and the metering pump. Mixer operation ranges from 30 minutes/day for the smallest system up to 55 minutes/day for the largest system. It is assumed that the chemical metering pump operates continuously 24 hours/day. Maintenance material costs are primarily for the mixer and metering pump, but maintenance material costs for a hydrated lime feed system are increased due to the continuous operation of the mixer.

Liquid Alum Feed--

Liquid alum feed systems are assumed to be custom-designed and contractor-constructed facilities for treatment systems in Categories 5 through 12. For smaller systems, the basic chemical feed system described above is satisfactory.

Conceptual Design--The design of these systems is based upon the use of liquid alum, which has a weight of 10 pounds/gallon. Fifteen days of storage are provided, using fiberglass reinforced polyester (FRP) tanks. The FRP tanks are assumed to be uncovered and located indoors for smaller installations, and outdoors for larger installations. Outdoor tanks are covered and vented, with insulation and heating provided to prevent crystallization, which occurs at temperatures below 30°F.

Operation and Maintenance Requirements--Electrical requirements are for feeder pump operation, building lighting, ventilation, heating, and heating needs of outdoor storage tanks. Maintenance materials include repair parts for pumps, motors, valves, and electrical starters and controls.

Labor requirements consist of time for chemical unloading and routine operation and maintenance of feeding equipment. Liquid alum unloading requirements are calculated on the basis of 1.5 hours/bulk truck delivery. Time for routine inspection and adjustment of feeders is 15 minutes/metering pump/shift. Maintenance requirements are assumed to be 8 hours/year for liquid metering pumps.

Polymer Feed--

Polymers are used as coagulants or filter aids. Polymer addition in treatment plants in Categories 1 through 4 is assumed to use a basic polymer feed system that is similar to the basic chemical feed system described above. Plants in Categories 5 through 12 use a polymer feed system like that described below.

Conceptual Design--The conceptual design of all polymer feed systems is based on preparation of a 0.25 percent stock solution.

The basic polymer feed system for Categories 1 through 4 consists of two tanks (a mixing tank and a storage tank), a mixer, and a metering pump. The mixing tank is elevated on a stand to just above the top water level in the storage tank to allow gravity flow of mixed polymer solution into the storage tank. The storage tank is the same size as the mixing tank, since the transfer of aged polymer solution to the storage tank should only be required once daily.

The larger polymer feed system for Categories 5 through 12 consists of a dry chemical storage hopper, a dry chemical feeder, two tanks, and a metering pump.

Operation and Maintenance Requirements--The operation labor requirements for the basic polymer feed system are primarily for daily measurement of dry polymer, adding it to the mixing tank, operating the mixer, and transferring the polymer solution to the storage tank. Additional labor is required for periodically checking the metering pump, preventive maintenance, and occasional repairs on the mixer and pump.

Operation and maintenance costs for larger polymer feed systems (Categories 5 through 12) are based on 3 percent of the manufactured equipment and pipe and valve costs. These costs do not include the cost of polymer. Labor requirements are for bag unloading (1 hour/ton of bags), the dry chemical feeder (110 hours/year for routine operation and 24 hours/year for maintenance), and the solution metering pump (55 hours/year for routine operation and 8 hours/year for maintenance).

Sodium Hydroxide Feed--

Sodium hydroxide is added for pH adjustment. Sodium hydroxide is in a dry form for plants in Categories 1 through 4 (up to 200 pounds/day). A basic chemical feed system can be used for this purpose. Plants in Categories 5 through 12 use liquid sodium hydroxide.

Conceptual Design--Dry sodium hydroxide (98.9 percent pure) is delivered to the plant site in drums and then mixed to a 10 percent solution on-site. A volumetric feeder is utilized to feed sodium hydroxide to the mixing tank. Two day-tanks are necessary: one for mixing a 10 percent solution, and one for feeding. The use of two tanks is necessary because of the slow rate of sodium hydroxide addition caused by the high heat of the solution. Each tank is equipped with a mixer, and a dual-head metering pump is used to convey the 10 percent solution to the point of application.

For feed rates greater than 200 pounds/day, a 50 percent sodium hydroxide solution is purchased premixed and delivered by bulk transport. The 50 percent solution contains 6.38 pounds of sodium hydroxide/gallon. For the 50 percent solution, 15 days of storage are provided in FRP tanks. Dual-head metering pumps are used to convey solution to the point of application, and a standby metering pump is provided in each case.

Pipe and valving is required for water conveyance to the dry sodium hydroxide mixing tanks and between the metering pumps and the point of application. The storage tanks are located indoors, since 50 percent sodium hydroxide begins to crystallize at temperatures less than 54°F.

Operation and Maintenance Requirements--Process energy requirements are for the volumetric feeder and mixer (smaller installations only) and the metering pump. A maintenance material requirement of 3 percent of equipment cost, excluding the storage tank cost, is utilized.

Labor requirements are based on unloading time for dry sodium hydroxide in drums, or the liquid 50 percent sodium hydroxide purchased in bulk for the larger installations. For installations using dry sodium hydroxide, additional labor required for routine operation time for the volumetric feeder is about 10 minutes/day feeder. In addition, for each installation, operation time for the dual-head metering pump is 15 minutes/day, with an annual maintenance time of 8 hours.

Lime Feed--

Lime, like sodium hydroxide, is added for pH adjustment. Lime feed systems are assumed to be used in plants in Categories 7 through 12.

Conceptual Design--Lime feed systems are assumed to use hydrated lime at feed rates up to 50 pounds/hour and quicklime at higher rates. Two types of feed systems are used: one that uses new lime, and one that uses a combination of new and recalcined lime. The use of recalcined lime lowers the storage requirement; therefore, decreases the cost of the lime feed facilities for a given capacity.

Hydrated lime is assumed to be purchased in 100-pound bags and to be fed using either a volumetric or gravimetric feeder to a slurry tank having a 5-minute detention time. In the slurry tank, the lime is mixed to a 6 percent slurry and fed by gravity to the point of application.

Quicklime is purchased in bulk, with 90 percent purity and a density of 60 pounds/ft^3, and stored in elevated hoppers above the lime slakers. Lime is conveyed pneumatically from a delivery truck to the storage hopper. Hoppers include a dust collector, bin gate, and flexible connection to the slaker. Hoppers are sized for a 30-day storage of lime, except when recalcination is used (then only a 3-day storage of the new lime is provided). The slaker mixes water with lime on a 2:1 weight basis, and the lime slurry is conveyed in an open gravity channel to the application point. Although all applications may not be able to use gravity conveyance in a channel, it is highly recommended from a maintenance standpoint. Standby slakers are included for all installations using quicklime.

Operation and Maintenance Requirements--Process energy requirements are for the feeder, slaker, and grit removal and are based on motor sizes for the various equipment. Annual maintenance material costs are based on 3 percent of manufactured equipment costs, excluding the cost of the storage hopper. Lime cost is not included in the maintenance material costs.

Labor requirements for unloading are 5 hours/50,000 pounds for bulk delivery and 8 hours/16,000 pounds for bag delivery and feed. Operation and maintenance time for the lime feeder, the slaker, and the associated grit removal is 3 hours/feeder/slaker/day. For smaller installations using hydrated lime, a routine operation time of 20 minutes/day is used for the solid feeder and the slurry tank.

Sulfuric Acid Feed--

Sodium hydroxide and lime are used to raise pH values. Sulfuric acid is used to lower pH. The feed system discussed in this report is capable of metering concentrated (93 percent) sulfuric acid from a storage tank directly to the point of application.

Conceptual Design--For sulfuric acid feed rates up to 200 gpd, the concentrated acid is delivered to the plant site in drums, and at larger flow rates it is delivered in bulk. Acid purchased in bulk is stored outdoors in FRP tanks, and acid purchased in drums is stored indoors. Fifteen days of sulfuric acid storage is provided, and a standby metering pump is included for all installations.

Operation and Maintenance Requirements--Process electrical energy requirements are for the metering pump. Building energy requirements are for indoor storage of the sulfuric acid drums. Maintenance material requirements were estimated at 3 percent of the equipment cost, excluding the cost of storage tanks. Labor requirements are for chemical unloading and for the metering pumps. Unloading times of 0.25 hour/drum of acid and 1.5 hours/bulk truck delivery were utilized. Metering pump routine operation is 15 minutes/pump/day, and maintenance requirements are 8 hours/feeder/year.

Filtration Process Components

The following paragraphs describe the conceptual design and operation and maintenance requirements of individual processes that are potential components of filtration treatment plants. Not all of the individual processes would be used in a single plant; rather, some would be used for direct filtration, others would be used for slow sand filtration, and so forth.

The processes described below are all related to treatment of the liquid stream in a treatment plant. Solids handling processes are described later in this section.

Rapid Mix--

Rapid mix units are used in this report both as part of a conventional treatment process train and with flocculation and clarification as a possible modification to a direct filtration plant. Costs of rapid mix units have therefore been developed for plants in Categories 5 through 12.

Conceptual Design--Construction costs are based on reinforced concrete basins sized for a 1-minute detention time at plant design capacity. The largest single basin capacity is 2,500 ft^3. Common wall construction is used when more than one basin is required. Mixer costs are for vertical shaft, variable-speed turbine mixers with 304 stainless steel shafts and paddler and totally enclosed fan-cooled EFC motors. Mixing energy is based on a G value of 900 sec^{-1}.

Operation amd Maintenance Requirements--Power requirements are based on the number of mixers, the G value of 900, a water temperature of 15°C, and an overall mechanism efficiency of 70 percent. Maintenance material costs consist of oil for the gearbox drive unit.

Labor requirements are determined using a jar testing time of 1 hour/day for plants under 50 mgd and 2 hours/day for plants over 50 mgd, 15 minutes/mixer/day for routine operation and maintenance, and 4 hours/mixer/6 months for oil changes. An allowance of 8 hours/basin/year is also included for draining, inspection, and cleaning.

Flocculation--

Flocculation basins are coupled with rapid mix units for conventional treatment in plants in Categories 5 through 12.

Conceptual Design--Flocculation basin costs are for rectangular, reinforced concrete structures 12 feet deep, with a detention time of 30 minutes at plant design flow. A length-to-width ratio of approximately 4:1 is used for basin sizing, and the maximum individual basin size utilized is 12,500 ft^3. Common wall construction is used where the total basin volume exceeded 12,500 ft^3. Costs are calculated for use of horizontal paddle flocculators, since horizontal paddles are less expensive for use in larger basins, and they generally provide more satisfactory operation in the larger basins, particularly when tapered flocculation is practiced.

A G value of 80 sec^{-1} is used to calculate manufactured equipment costs. All drive units are variable-speed to allow maximum flexibility. Although common drive for two or more parallel basins is commonly utilized, the estimated costs were calculated using individual drive for each basin.

Operation and Maintenance Requirements--Energy requirements are calculated on the basis of a horsepower per unit volume requirement of 0.17 foot-pounds/second/ ft^3. An overall motor/mechanism efficiency of 60 percent is utilized.

Maintenance material costs are based on 3 percent of the manufactured equipment costs. Labor requirements are based on routine operation and maintenance of 15 minutes/day/basin (maximum basin volume = 12,500 ft^3) and an oil change every 6 months requiring 4 hours/change. No allowance is included for jar test time, as this is included in the rapid mix operation and maintenance curves.

Rectangular Clarifiers--

Rectangular clarifiers are used in process groups following rapid mix and flocculation in Categories 5 through 12.

Conceptual Design--Clarifiers are sized using an overflow rate of 1,000 gpd/ft^2 and a minimum of two basins. Cost estimates are made for clarifiers that have a 12-foot sidewall depth and use chain and flight sludge collectors.

The construction costs include the chain and flight collector, the collector drive mechanism, weirs, the reinforced concrete structure complete with inlet and outlet troughs, a sludge sump, and sludge withdrawal piping. Costs for the structures are developed assuming multiple units with common wall construction. Yard piping to and from the clarifier is not included.

Operation and Maintenance Requirements--Process energy requirements are calculated based on manufacturers' estimates of motor size and torque requirements for lime sludges.

Maintenance material costs are for parts required for periodic maintenance of the drive mechanism and weirs. Labor requirements are for periodic checking of the clarifier drive mechanism, as well as periodic maintenance of the mechanism and weirs.

Tube Settling Modules--

Tube settling modules employing principles of shallow-depth sedimentation can be incorporated in clarification basins to significantly reduce the size and associated construction cost of new basins. The modules may also be used to improve performance or increase capacity at existing facilities. Separate costs for adding tube settling modules to an existing clarifier are included in this report.

Conceptual Design--The conceptual design for the modules is based on a rise rate of 2.0 gpm/ft^2 through the area covered by the modules. By leaving a portion of the basin open, a zone is created for inlet turbulence dissipation. This transition zone is separated by a baffle extending from the bottom of the modules to 6 inches above the operating water level. Uniform effluent collection is a requirement for optimum utilization of tube settlers. To meet this requirement, effluent launders are spaced at 12-foot centers in all basins. Since the hydraulic and structural requirements for tube clarification systems are unique, the costs include tube modules, tube module supports and anchor brackets, a transition baffle, effluent launders with V-notch weir plates, and installation. These costs may be added to the conventional basin construction costs to arrive at a total facility cost.

Gravity Filtration--

The gravity filtration process actually includes two components: the filtration structure and the filtration media. For purposes of this report, the filtration structures are sized on the basis of an application rate of 5 gpm/ft^2 using mixed media. Gravity filtration costs are included in the costs for conventional treatment for Categories 5 through 12.

Conceptual Design--Conventional gravity filtration structure costs are based on use of cast-in-place concrete with a media depth of 2 to 3 feet and a total depth of 16 feet for the filter box. At flows less than 5 mgd, two filters are used, but at 5 mgd and greater, a minimum of four filters are utilized. Maximum filter size is limited to 1,275 ft^2, and above 700 ft^2, the filters are dual-celled to allow backwashing of each half separately. This approach allows a significant reduction in the size of washwater and waste piping. Costs for filtration structures include the filter structure, underdrains, washwater troughs, a pipe gallery, required piping and cylinder-operated butterfly valves, filter flow and headloss instrumentation, a filter control panel, and housing of the entire filter structure.

Construction costs for mixed media are based on the purchase and placement of 30 inches of media over a 12-inch gravel underdrain. These estimates are applicable to either gravity or pressure filters, although pressure filters are often designed with a somewhat deeper gravel support layer. For plants with total filter areas of 140 ft^2 and less, materials are assumed to be truck shipped in 100-pound bags. For total filter areas between 140 and 2,000 ft^2, rail shipment in 100-pound bags is assumed, and for larger filter areas, rail shipment by bulk is assumed. The estimated costs include media cost, shipping, installation, and the cost of a trained technician to direct placement of the media.

Operation and Maintenance Requirements--Energy requirements are only for building heating, ventilation, and lighting. All process energy required for filtration is included in the backwash and surface wash process costs.

Maintenance material costs include the costs of general supplies, instrumentation repair, and the periodic addition of filter media. Labor costs include the cost of operation, as well as the cost of instrument and equipment repairs and supervision.

Convert Rapid-Sand Filters to Mixed-Media Filters--

The cost for this conversion simply assumes the exchange of rapid-sand media for mixed media in an existing filtration structure. No changes are made in influent or effluent piping, or in flow control devices. Costs include removal of sand media and purchase and placement of 30 inches of mixed media over a gravel underdrain.

Filter-to-Waste Facilities--

As indicated in Section III of this document, filter-to-waste facilities are recommended for slow-sand filtration and may be advisable for other filtration systems. The piping, fittings, and valves included in these facilities are sized to convey filtered water at an application rate of 5 gpm/ft^2.

Capping Sand Filters with Anthracite--

A popular technique for increasing the capacity of existing rapid-sand filter installations involves removing the top 6 to 12 inches of sand and replacing it with anthracite coal. The coarser coal permits suspended solids penetration into the filter bed, allowing operation of the filter beds at higher flow rates and for longer periods between backwash. In many situations, this modification can effect a 30 to 50 percent increase in capacity and a reduction in washwater usage.

Conceptual Design--Costs are developed assuming the removal of 12 inches of sand and replacement with 12 inches of anthracite coal. The costs include labor for removing the sand from the filter and disposing it on-site, material and freight costs for anthracite coal, and installation labor. Labor costs assume that sand removal from filters smaller than 3,500 ft^2 would be accomplished by manual labor. For larger filters, manual labor is supplemented with mechanical equipment.

Slow-Sand Filters--

The slow-sand filter systems in this report are based on use of a cast-in-place, concrete-covered structure for use in Categories 1 through 3 and uncovered earthen berm structures for Categories 4 through 7. Uncovered earthen berm structures have significantly lower construction costs at larger plant sizes. Concrete-covered structures are cost competitive at small plant sizes, have no freezing problems during winter months, no algal growths in the water above the filter, and no problems with windblown debris or bird droppings.

Conceptual Design--Slow-sand filter systems are sized on the basis of a filtration rate of 70 gpd/ft^2 (3.0 mgd/acre). The depth of the sand bed is 3.5 feet over 1.0 feet of support gravel. A supernatant box height of 4.0 feet is provided above the filter surface.

Covered filters incorporate below-grade concrete structures with beam and slab covers. All piping is either cast iron or steel and is also below grade. Because slow-sand filters operate most efficiently at a constant filtration rate, flowmeters for each filter and a below-grade, concrete clearwater reservoir are included. The rate of flow through the filter is established by an individual control valve for each filter. Also included in the concept is an effluent flow control structure. This structure has two principal purposes: (1) to prevent the water level in the sand filter from inadvertently being drawn down below the filter surface; and (2) to aerate the filter effluent to release any odors that may be present.

Uncovered, slow-sand filters include on-grade, clay-lined structures formed by earthen berms. All piping is PVC and is below grade. A steel tank reservoir, an effluent flow control structure, and a flowmeter to measure total plant flow are also included.

Operation and Maintenance Requirements--Process energy requirements are negligible. Building energy requirements are for heating, lighting, and ventilation of the control and storage building. Fuel is required for the mechanical sand bed cleaner.

Maintenance material requirements are based upon anticipated costs of replacement parts and replenishment of the minor amount of consumable supplies involved in daily operation. The cost of replacement sand is not included. Experience with operating filters suggests that careful cleaning allows resanding to be necessary as infrequently as every 10 years.

Labor requirements are developed assuming the filters operate unattended. They include a short daily inspection and control valve adjustment, daily turbidity measurements, quarterly cleanings of each filter by hand, normal repair and replacement of equipment.

Pressure Filtration--

Surface water treatment by direct filtration using pressure filters is one of the modes of direct filtration for which costs are presented in this report. Pressure filtration is most commonly used for treatment in plants in Categories 3 through 10.

Conceptual Design--Pressure filter systems in this report include mixed media and are sized on a filtration rate of 5 gpm/ft^2 using 24-inch to 36-inch deep filter beds.

Cost estimates are based on use of either vertical or horizontal cylindrical ASME code pressure vessels of 50 to 75 psi working pressure. Each plant consists of a minimum of four vessels. Filter vessels are provided with a pipe lateral underdrain, and filter media are supported by graded gravel. This type of underdrain is suitable for water backwash with surface wash assist. For air-water backwashing, the pipe laterals are replaced with a nozzle underdrain.

Costs include a complete filtration plant with vessels, cylinder-operated butterfly valves, filter face piping and headers within the filter gallery, filter flow control and measurement instrumentation, headloss instrumentation, and a master filter control panel. The filters are designed to backwash automatically on an input signal such as headloss, turbidity breakthrough, or elapsed time, or by manual activation. Not included in the cost estimate are supply piping to the filtration units from other unit processes, filter supply pumping, backwash storage and pumping, surface wash or airwash supply facilities, or filtration media.

Operation and Maintenance Requirements--Energy requirements are based on the conceptual designs for process, and for heating, lighting, and ventilating. Process energy is for the filtration system supply pumps and backwash pumps. Continuous 24 hour/day, 365 day/year operation with one backwash/day of a 10-minute duration is assumed. It is further assumed that the surface wash supply will be obtained from the pressurized distribution system with suitable means for backflow prevention.

Maintenance material costs are for additional filter media, charts and ink for recorders, and miscellaneous repair items for electrical control equipment and valves. Labor requirements are based on records from operating plants.

Contact Basins for Direct Filtration--

Contact basins are often used for pretreatment ahead of gravity or pressure filters in direct filtration applications, as described in Section III of this report. Costs are developed for contact basins both as part of a gravity-filter direct filtration plant, and as a separate structure that could be added to an existing plant in Categories 4 through 12.

Conceptual Design--Costs are based on construction of open, reinforced concrete contact basins with a 1-hour detention time. The basins are similar to rectangular clarifiers, with an 11-foot water depth.

Operation and Maintenance Requirements--Since these basins are not equipped with sludge collection mechanisms, seasonal draining and cleaning is the only operation and maintenance requirement.

Hydraulic Surface Wash Systems--
If surface wash is utilized, the cost must be added to the costs of the backwash system, the filter structure, the filter media, and any required backwash storage capacity to arrive at the total cost of filtration.

Conceptual Design--Cost estimates include dual pumps with one as standby, electrical control, piping, valves, and headers within the filter pipe gallery. No allowance for housing is included, as this is included in the filtration structure cost. Surface wash pumps are sized to provide approximately 50 to 85 psi at the arms.

Costs are based on the total filter area of a plant, using the filter conceptual designs presented in the gravity filtration structure section. One dual-arm agitator is included for filter areas up to and including 75 ft^2, four agitators for 350 to 700 ft^2 filters, and six and eight agitators for 1,000 ft^2 and 1,275 ft^2 filters, respectively. It is assumed that the wetwell for the surface wash pumps is the same as for the backwash pumps.

Operation and Maintenance Requirements--Energy requirements are based on two surface washes/day, a wash time of 8 minutes, an application rate 1.5 gpm/ft^2 of filter surface, a TDH of 200 feet, and an overall motor pump efficiency of 70 percent.

Maintenance material requirements are for repair of the pump(s), motor starter, valves and surface agitators. Labor requirements are for maintenance of equipment only. Operation labor is included with the basic filter.

Washwater Surge Basins--
When filter washwater is recycled through the plant, treated in a separate plant, or discharged to a sanitary sewer, a surge basin may be used to even out the flow.

Conceptual Design--Surge basins are included in conceptual designs of gravity and pressure filtration systems for Categories 3 through 12. Construction costs are based on covered, underground, reinforced concrete basins with level control instrumentation. The basins are sized to store the backwash flow from one filter for 20 minutes at a flow rate of 18 gpm/ft^2.

Automatic Backwashing Filter--
The continuous automatic backwash filter is an adaptation of rapid-sand filtration principles. The filter bed is contained in a shallow rectangular concrete structure that is laterally divided into compartments. Each compartment is, in effect, a single filter. Filter flow rate is based on declining rate, as there are no rate-of-flow controllers. An attractive feature of the filter is that operating headlosses are generally less than 1 foot of water. A motor-driven carriage assembly equipped with a backwash pump and a washwater collection pump backwashes each compartment sequentially as it traverses the length of the filter.

Conceptual Design--Costs are based on a filtration rate of 3 gpm/ft^2. A filter box depth of 5 feet is used for all sizes of filters, and each plant size utilizes a minimum of two filters. The filter units are essentially self-contained and require no interconnecting piping. Filtered water, influent, and backwash water are conducted to and from the filter by troughs or channels integrally cast within the concrete filter structure.

The nature of the equipment requires that it be housed for protection from inclement weather and freezing temperatures. Housing costs assume total enclosure of the filters with minimum space for access on two of the four sides for maintenance.

Construction costs include the filtration structure, internal mechanical equipment, partitions, underdrains, rapid-sand filter media (depth generally 11 inches) washwater collection trough, over-head pump carriage, electrical controls, and instrumentation.

Operation and Maintenance Requirements--Energy requirements are for building heating, lighting, and ventilation, and for pumping costs related to backwashing. Maintenance material costs include general supplies, pump maintenance and repair parts, replacement sand, and other miscellaneous items. Labor costs are estimated from projected maintenance time requirements and are related to general supervision and maintenance.

Clearwell Storage--
Treated water is commonly stored at the plant site before pumping, as a supplement to distribution system storage. In many cases, filter washwater pumps also draw from the clearwell, eliminating the need for a separate sump. Clearwell storage is included in surface water treatment plants of all sizes.

Conceptual Design--Storage volumes are sized to contain between 5 percent and 15 percent of treatment design capacity for larger to smaller plants, respectively.

Clearwell storage may be below-ground in reinforced concrete structures, or aboveground in steel tanks. Instrumentation and control of the clearwell water level is very important in terms of pacing the plant output. In addition, instrumentation for turbidity and chlorine measurement, as well as other quality control operations, is normally provided with the clearwell.

Package Pressure Diatomite Filtration--
A pressure diatomite filter passes pressurized water through a layer of diatomaceous earth filter aid for the removal of suspended solids. Package filters are practical for water treatment in Categories 1 through 4.

Conceptual Design--A typical package system is sized on the basis of 1 gpm/ft^2, and consists of:

- Diatomite storage, preparation, and feeding equipment
- A pressure filtration vessel
- A filter supply pump
- Valves, piping, and fittings
- A control panel
- Housing

Construction costs are for a complete installation including diatomaceous earth storage, preparation and feed facilities, pressure filtration units, a filter supply pump, filter valves, interconnecting pipe and fittings, and a control panel for automatic operation. Housing costs are for enclosure of the filters in a modular steel building, with minimum space for access on all sides of the filter units for maintenance.

Operation and Maintenance Requirements--Process energy requirements are for filter feed pumps, a body feed pump, mixers, and other items associated with the filter system. A cycle time of 24 hours between cake removals is assumed in developing electrical requirements. The energy requirements do not include those associated with raw water or finished water pumping.

Maintenance material requirements are related primarily to replacement of pump seals, application of lubricants, chemical feed pump replacement parts, and general facility maintenance supplies.

Labor requirements are developed assuming that the diatomite filter installation operates automatically. Operator attention is necessary only for preparation of body feed and precoat, and for verification that chemical dosages are proper and that the equipment is producing a satisfactory quality filtered water.

Chemical Requirements--A "typical" chemical dosage is difficult to establish because of the widely varying characteristics of the raw water supplies. For raw water turbidities of less than 5 NTU (recommended upper limit for application of diatomite filtration), filter aid dosages commonly range between 10 to 30 mg/L, including both precoat and body feed.

Pressure Diatomite Filters--

Costs for diatomaceous earth filtration in plant Categories 5 through 12 are based on use of pressure diatomite filters as one of several separately housed processes.

Conceptual Design--Pressure diatomite filters are based on a filter loading rate of 1.6 gpm/ft^2, and include all the facilities described above for package pressure diatomite filters.

Operation and Maintenance Requirements--Process energy usage is for filter pumps, backwash pumps, mixers, and other items associated with the filter system. A cycle time of 24 hours between backwashes is assumed in developing electrical requirements. The energy requirements do not include energy associated with raw water or finished water pumping.

Maintenance material requirements include replacement of pump seals, application of lubricants, instrument and chemical feed pump replacement parts, and general facility maintenance supplies. Costs for treatment chemicals include diatomaceous earth and other necessary chemicals.

Labor requirements assume that the diatomite filter installation operates automatically and virtually unattended. Operator attention is only necessary for preparation of body feed and precoat and for verification that chemical dosages are proper and that the equipment is producing a quality filtered water.

Package Ultrafiltration Plants--

Ultrafiltration utilizes a hollow fiber membrane to remove undissolved, suspended, or emulsified solids from water. Package ultrafiltration plant costs are included in this document for surface water treatment plants in Categories 1 through 4.

Conceptual Design--A typical package ultrafiltration plant consists of:

- Hollow fiber membrane cartridges
- Automatic and manual valves for backwashing and unit isolation
- Flowmeters
- Pressure gauges
- A backwash pump
- A control panel
- A supply pump
- Membrane cleaning solution pumps and storage tanks
- Package plant housing

Product water storage facilities are not included. Ultrafiltration plants are designed for an initial flux rate of approximately 70 gpd/ft^2.

Operation and Maintenance Requirements--Process energy requirements are calculated using connected horsepower sizes recommended by manufacturers. Continuous 24 hour/day, 365 day/year operation with one backwash/day of a 30-minute duration is assumed in the process energy calculations.

Maintenance material requirements include replacement of hollow fiber membrane cartridges once every 4 years, replacement of pump seals, small parts for chemical feed pumps and instruments, and for general facility operation.

Chemical Requirements--Chlorine is the only chemical used routinely in ultrafiltration systems. Chlorine is added to the backwash water at dosages ranging up to 50 mg/L to assist in cleaning the membranes.

Package Conventional Complete Treatment--
Costs for use of package water treatment plants are developed for surface water treatment plants in Categories 1 through 6. These units include coagulation, flocculation, sedimentation, and filtration, but exclude raw water pumping, clearwell storage, and finished water pumping.

Conceptual Design--Cost estimates are for standard manufactured units incorporating 20 minutes of flocculation, tube settlers rated at 150 gpd/ft^2, mixed-media filters rated at 2 gpm/ft^2, and a media depth of 30 inches. The costs include premanufactured treatment plant components, mixed media, chemical feed facilities (storage tanks and feed pumps), flow measurement and control devices, pneumatic air supply (for plants of 200 gpm and larger) for valve and instrument operation, effluent and backwash pumps, and all necessary controls for a complete and operable unit and building. Smaller plants (below 50 gpm) utilize low-head filter effluent transfer pumps and are used with an above-grade clearwell. Larger plants gravity discharge to a below-grade clearwell.

Operation and Maintenance Requirements--Process energy requirements are for chemical mixers, chemical feed pumps, mechanical flocculators, backwash pumping, and where applicable, surface wash pumping. Chemical feed pumps are assumed to operate 24 hours/day, while chemical mixers are only operated for 15 to 30 minutes following preparation of a new solution. Backwash is assumed to last for 7 minutes at a rate of 15 gpm/ft^2. Larger plant sizes are equipped with surface wash rated at 50 psi and operate for 4 minutes/backwash.

Building energy is required for heating, ventilating and lighting, with the latter assumed to be 3 hours/day. Maintenance materials are for replacement of mechanical and electrical components that wear out or break down during normal operation.

Labor requirements are for maintenance and operation, with the majority of labor being required for operation. The operator is required, on a daily basis, to maintain chemical storage tanks filled with chemicals in the proper concentrations, to make laboratory measurements for determination of chemical dosage rates, to assure correct chemical feed rates, and to perform other routine laboratory testing as required. Labor requirements increase when the filtration rate increases since more labor is required for chemical preparation. Some maintenance items may also be required on a daily basis, as will general housekeeping.

Chemical Requirements--Chemicals most commonly used in package conventional plants include alum, polymer, chlorine, and soda ash.

Disinfection Processes

While some water supply systems may not require filtration or the filtration-related processes described in the preceding pages, all will require disinfection by one or more of the processes described in Section IV. The following paragraphs present the conceptual designs and operation and maintenance requirements for the disinfection processes for which costs are tabulated at the end of this section.

Chlorine Storage and Feed Systems--

The costs for chlorine feed facilities are based on use of hypochlorite solution feed for Categories 1, 2, and 3; 150-pound cylinders for feed rates up to 100 pounds/day; ton cylinders for feed rates up to 2,000 pounds/day; and on-site storage with bulk rail delivery for rates above 2,000 pounds/day.

Conceptual Design--The hypochlorite solution feed process is described in Section V. For cylinder storage, maximum chlorinator capacity is 8,000 pounds/day, and one standby chlorinator is included for each installation. Construction costs include cylinder scales and evaporators for delivery rates of 2,000 pounds/day and greater. Residual analyzers with flow-proportioning controls are included for flow rates greater than 1,000 pounds/day. Costs are included for injector pumps capable of delivering sufficient water at 25 psi to allow production of a 3,500-mg/L solution. Housing costs include the chlorinator room and 30 days of cylinder storage. For feed rates greater than 100 pounds/day, electrically operated, monorail trolley hoists are also included.

Use of an on-site storage tank eliminates the housing requirement for cylinder storage, the monorail and hoist, the cylinder scale, cylinder trunnions, and the cylinder manifold piping. However, other costs are incurred for the tank and its supports, a tank sun shield, load cells for the tank, a railhead connection and associated track, an unloading platform, an air padding system, expansion tanks, and miscellaneous gauges, switches, and piping. All considerations relating to the chlorinators, evaporators, and other feed equipment remain the same as for the ton cylinder system. The amount of chlorine storage provided with the on-site tank is 30 days.

The rail siding includes the cost of a turnout from the main line, 500 feet of on-site track, and the unloading platform. Piping costs are strongly influenced by the location of the storage tank relative to the chlorinators. Normally, the storage tank is located near the plant boundary. Valving is more complex than with ton cylinders, mainly because of the unloading system, the use of duplicate heads for gas or liquid feed, and the air padding system.

Operation and Maintenance Requirements--Power requirements include heating, lighting, and ventilation of the chlorination building and the cylinder storage area, the electrical hoist when ton cylinders are used, evaporators when feed is 2,000 pounds/day or greater, and the injector pump for the high-strength chlorine solution. This pump was sized to deliver sufficient flow for a maximum chlorine concentration of 3,500 mg/L in the high-strength solution. Where on-site storage tanks are utilized, the electrical hoist power requirements are eliminated, and heating, ventilating, and lighting power are significantly reduced as a result of elimination of indoor storage facilities.

Maintenance material requirements are based on experience at operating plants and are essentially the same for use of cylinders or on-site storage. Cost of chlorine is not included in the maintenance material estimates, but is included in chemical costs.

Labor requirements for cylinders are based on loading and unloading cylinders from a delivery truck, time to connect and disconnect cylinders from the chlorine headers, and the time for routine daily checking of the cylinders. For on-site tank storage, labor consists of time to unload a bulk delivery truck or rail tanker. Common to all installations would be the time required for daily checking and periodic maintenance of the chlorine handling system.

Chlorine Dioxide Generation and Feed--
Chlorine dioxide is most commonly generated by mixing a high-strength chlorine solution with a high-strength sodium chlorite solution in a PVC chamber filled with procelain rings. Chlorine dioxide may also be generated by acidifying solutions of sodium chlorite and sodium hypochlorite with sulfuric acid. This method is only applicable in very small installations with little operator time available, and is not included in the costs in this document.

Conceptual Design--In theory, 1.34 pounds of pure sodium chlorite and 0.5 pounds of chlorine react to give 1 pound of chlorine dioxide. However, since sodium chlorite is normally purchased with a purity of 80 percent, 1.68 pounds of sodium chlorite are required per pound of chlorine dioxide generated. Chlorine is normally used at a 1:1 ratio with sodium chlorite to ensure completion of the reaction and to lower the pH to 4. Costs are based on using 1.68 pounds of chlorine and 1.68 pounds of sodium chlorite per pound of chlorine dioxide generated. Costs include a sodium chlorite mixing and metering system, chlorine dioxide generator, polyethylene day tank, mixer for the day tank, and dual-head metering pump. The chlorine dioxide generator is sized for a detention time of about 0.2 minutes.

Operation and Maintenance Requirements--Electrical requirements include power for the gaseous chlorination system, the sodium chlorite mixing and metering system, and building heating, lighting, and ventilation.

Maintenance material requirements are based on experience with gaseous chlorine systems and liquid metering systems. Costs for sodium chlorite and chlorine are included as chemical costs. Labor requirements consist of labor for gaseous chlorination systems, plus the labor required to mix the sodium chlorite solution, to adjust its feed rate, and to maintain the mixing and metering equipment.

Ozone Generation, Feed, and Contact Chambers--

Ozone may be generated on-site using either air or pure oxygen. For systems up to 100 pounds/day, air was assumed to be the feed. At generation rates greater than 100 pounds/day, pure oxygen generated on-site is the feed for the ozone generator.

Conceptual Design--The manufactured equipment cost for ozone generation includes the gas preparation equipment, oxygen generation equipment (at more than 100 pounds/day), the ozone generator, dissolution equipment, off-gas recycling equipment, electrical and instrumentation costs, and all required safety and monitoring equipment. All ozone-generating equipment is considered to be housed, but all oxygen-generating equipment is located outside on a concrete slab.

The ozone contact chamber is a covered, reinforced concrete structure with a depth of 18 feet, and a length-to-width ratio of approximately 2:1. Partitions are utilized within the chamber to assure uniform flow distribution. Ozone dissolution equipment costs are included within the ozone generation cost curve and are not included with the ozone contact chamber.

Operation and Maintenance Requirements--For ozone generation systems, electrical energy is required for the ozone generator and building heating, cooling, and lighting requirements. Ozone generation using air feed, generally in systems producing less than 100 pounds/day, requires 11 kWh/pound of ozone generated. For larger, oxygen-fed systems, the power requirements are 7.5 kWh/pound of ozone generated. These figures include oxygen generation, ozone generation, and ozone dissolution.

Maintenance material requirements are for periodic equipment repair and replacement of parts. Labor requirements are for periodic cleaning of the ozone generating apparatus, maintenance of the oxygen generation equipment, annual maintenance of the contact basin, and day-to-day operation of the generation equipment.

Ammonia Feed Facilities--
Ammonia feed facilities are required when chloramination is used as the method of disinfection. Ammonia may be fed in either of two forms--anhydrous ammonia or aqua ammonia. Anhydrous ammonia is purchased as a pressurized liquid and is fed through evaporators and ammoniators and then as a gas to the point of application. Aqua ammonia is a solution of ammonia and water that contains 29.4 percent ammonia. Aqua ammonia is metered as a liquid directly to the point of application.

Costs in this document are based on the use of anhydrous ammonia up to a feed rate of 100 pounds/day, and aqua ammonia facilities for feed rates greater than 100 pounds/day.

Conceptual Design--For anhydrous ammonia, construction costs include bulk ammonia storage for all feed rates, with 10 days of storage provided. The storage tanks include the tank and its supports, a scale, an air padding system, and all required gauges and switches. The ammonia feed system consists of an ammoniator and flow-proportioning equipment. Dry ammonia gas is fed directly to the point of application.

When aqua ammonia is used, it is stored in a horizontal pressure vessel with a length/width ratio of approximately 3:1. One tank is used for each installation, and the usable storage capacity is 10 days. Construction costs include the tank and its supports, and required piping and valving to fill the tank from a bulk delivery truck and to convey aqua ammonia from the tank to the metering pump and then to the point of application. A housing cost is not included because only the metering pump is housed, and it could easily be located in a number of other housed areas.

Operation and Maintenance Requirements--For anhydrous ammonia feed facilities, electrical energy requirements are for heating, lighting, and ventilation of the ammoniator building.

Maintenance material requirements are based on operating experience at chlorination facilities of similar size. Anhydrous ammonia costs are included as chemical costs. Labor requirements are for transfer of bulk anhydrous ammonia from a delivery truck or rail car to the on-site ammonia storage tank, plus day-to-day operation and maintenance requirements. A bulk unloading time of 3 hours/shipment is utilized. Operation and maintenance requirements vary from roughly 1.5 hours/day for the smaller systems to 3 hours/day for larger systems.

For aqua ammonia feed facilities, electrical energy costs are only for operation of the metering pump. Because of the small indoor area required for the metering pump and standby pump, no allowance is included for building heating, lighting, and ventilation. Transfer of aqua ammonia from bulk truck to the storage tank is assumed to be by a pump located on the bulk truck.

Maintenance material costs are for repair parts for the metering pump, valve repair, and painting of the storage tank. Aqua ammonia costs are included as chemical costs. Labor costs include 15 minutes/day for operational labor, 24 hours/year for maintenance labor, and 1 hour/unloading of the bulk delivery truck.

Ultraviolet Light Disinfection--
Costs of disinfection with ultraviolet light have been developed for groundwater treatment Categories 1 through 4.

Conceptual Design--Construction costs include single and multiple ultraviolet units ranging in capacity from 10 to 780 gpm. The units are furnished in modular form requiring only piping and electrical connections. The units are compact (a 780-gpm module would occupy an area of less than 24 ft^2). The costs include the manufactured units and related piping, electrical equipment, equipment installation, and a building to house the equipment.

Operation and Maintenance Requirements--Process energy is for a mercury lamp in continuous 24 hours/day operation with only occasional shutdown to clean cells and replace weak ultraviolet lamps. Building energy is for heating, lighting, and ventilation. Maintenance materials are related to the replacement of the ultraviolet lamps, generally after operating continuously for about 8,000 hours. Labor requirements are related to occasional cleaning of the quartz sleeves and periodic replacement of the ultraviolet lights.

Solids Handling Processes

While not a major topic in this document, solids handling processes must be a part of any water treatment plant designed to remove turbidity and suspended solids. For simplicity in preparing the costs at the end of this section, only six solids handling processes are used, depending on the size and type of treatment plant. As described below the processes include sludge holding tanks, sludge dewatering lagoons, liquid sludge hauling, gravity sludge thickeners, filter presses, and dewatered sludge hauling.

Sludge Holding Tanks--

Sludge holding tanks are normally used to dampen fluctuations in the quantity and quality of sludge flow to the thickening and dewatering facilities. In some cases they are also used to blend sludge from various sources.

Conceptual Design--Construction costs are mainly a function of tank volume, but also include mixers for the tanks. Sludge pumping to or from the holding tanks are not included.

Operation and Maintenance Requirements--Operation and maintenance requirements include labor and materials for maintaining the tank mixer and for periodic cleaning of the tanks. Process energy is required for the mixer.

Sludge Dewatering Lagoons--

Sludge dewatering lagoons are sized to contain a 1-year volume of solids, and have been used for treatment plants in Categories 1 through 7.

Conceptual Design--Construction costs are for unlined lagoons with 2-foot freeboard depth and 3:1 side slopes. Water depth is 7 feet in basins less than 40,000 ft^3 volume and 10 feet for larger volumes. It is assumed that the excavation volume is equal to the dike fill volume. Lagoons include an inlet structure that would prevent disturbance of settling material, and an outlet structure to skim clarified water.

Operation and Maintenance Requirements--Operation and maintenance requirements are primarily associated with sludge removal from the lagoons. Depending on the climate and the ability of water to percolate from the lagoon, sludge can thicken to a solids content of 20 to 50 percent during 6 months of storage. Removal is generally done with a front-end loader or with dragline dredging. Dredging is used to allow further dewatering by air drying on the lagoon periphery. After air drying, the concentrated sludge is removed by a front-end loader. The costs and requirements presented are for a combination of these approaches. Sludge is assumed to be removed from a lagoon on the average of once every 2 years, and hauled in dump trucks to within 1 mile of the lagoons.

Energy requirements are for diesel fuel used in the removal and transport of sludge. Maintenance material is for periodic lagoon grading and restoration of dikes and roadway maintenance. Labor requirements are for sludge removal and transport, and for lagoon maintenance.

Liquid Sludge Hauling--

Liquid sludge hauling costs are based on an assumed 5 percent solids concentration.

Conceptual Design--Costs are for hauling liquid sludge in agency-owned 5,500-gallon tanker trucks with volumes ranging between 1.25 and 100 million gallons/year, for a one-way distance of 20 miles. Loading facilities include a truck-loading enclosure and appropriate piping and valving to allow loading in a maximum time of 20 minutes. Pumping facilities are not included, but can be added separately by chemical sludge pumping. Truck usage is assumed to be 8 hours/day.

Operation and Maintenance Requirements--Energy requirements for sludge hauling are for diesel fuel. Maintenance items include tires, normal maintenance, lubricants, and vehicle insurance. Labor requirements are for driver time and for time associated with loading and unloading the tanker truck.

Gravity Sludge Thickeners--

Gravity sludge thickeners are used in treatment plant Categories 8 through 12 for purposes of developing cost estimates. They are sized on the basis of 100 gpd/ft.

Conceptual Design--Gravity thickeners are similar to circular clarifiers, although the thickener mechanism is somewhat different than a sludge scraper in a clarifier. Construction costs include the thickener mechanism and its installation, the reinforced concrete structure with 12-foot side wall depth, and concrete effluent troughs with the weir baffle located on the inboard side of the trough. All mechanisms are supported by the center column and include a steel bridge from the edge of the thickener to the center column.

Operation and Maintenance Requirements--Process energy requirements are for driving the thickener mechanism. Maintenance material costs and labor requirements are for repair and normal maintenance of the thickener drive mechanism and weirs.

Filter Press--

Sludge dewatering by filter presses has been used, together with gravity sludge thickening, to develop estimated surface water treatment costs for Categories 8 through 12. Filter presses have been sized on the basis of a feed rate of 5 pounds (dry) solids/ft /hour, assuming a 19 hour/day operation.

Conceptual Design--The construction costs include the filter press, feed pumps (including one standby), a lime storage bin and feeders, a sludge conditioning and mixing tank, an acid wash system, and housing.

Operation and Maintenance Requirements--Most of the process energy consumed by the filter press is related to operation of the feed pump. Energy is also consumed by the open-close mechanism and the tray mover. Pumping power requirements are calculated for a solids loading of 4 percent at a cycle time of 2.25 hours, with a 0.33-hour turnaround time between cycles. Power required for chemical preparation, mixing, and feeding is also included in process energy. Energy requirements related to building heating, lighting, and ventilation are also included.

Dewatered Sludge Hauling--
Hauling of dewatered sludge is used in cost estimates for surface water treatment in Categories 8 through 12, following gravity sludge thickening and dewatering by filter presses.

Conceptual Design--Costs are for hauling dewatered sludge in agency-owned trucks over a one-way distance of 20 miles. Loading facilities include a sludge conveyor, a hopper capable of holding 1.5 truckloads of sludge, and an enclosure for the sludge hopper. When more than one hopper is required, multiple conveyors and enclosures are utilized.

Operation and Maintenance Requirements--Energy requirements for sludge hauling are for diesel fuel. Process energy for sludge pumping at the treatment facility is not included. Maintenance costs for the trucks are calculated on a $/mile-traveled basis, and do not include fuel.

Labor requirements are for the truck operators. A loading time of 20 minutes and an unloading time of 15 minutes are utilized, and it is assumed that the truck operator would be responsible for each.

Administration, Laboratory and Maintenance Building

For purposes of this document, filtration plants in Category 5 or above are assumed to have separate building area(s) set aside for administrative offices, laboratory, maintenance or shop area, and general storage. In smaller plants, the areas are assumed to be either incorporated into other plant housing structures, or located completely off the plant site and are not a part of treatment costs.

Conceptual Design--Building areas are specifically for the purposes stated above and do not include area for chemical storage, which is included in the design of the individual feed systems. Costs for these building areas also include such items an laboratory equipment or supplies, maintenance equipment, and vehicles.

Operation and Maintenance Requirements--Building energy requirements are based on area requirements obtained from more then a dozen water treatment plants.[1] Maintenance material costs are related to operation and maintenance of administrative facilities that are not directly assignable to specific plant components. Such expenses include office supplies, communications, dues, subscriptions, office equipment repairs, travel expenses, training course expense, and custodial supplies.

Labor requirements are limited to personnel whose time and effort are related only to administration and management of the plant, such as superintendent, assistant superintendent, plant chemist, bacteriologist, clerk, and maintenance supervisor. Note that operation and maintenance labor, which is included under each specific unit process, is excluded from these costs.

ALTERNATIVES TO TREATMENT

The cost to treat a surface water supply may be higher than the cost of an acceptable alternative that does not involve treatment. A prime example of such an alternative would be to find or provide a new water source. Two methods of developing an alternative, or supplemental, supply source are described below. Both would be more practical alternatives for small water systems.

Constructing a New Well

Water wells are drilled by the cable tool, hydraulic rotary, or reverse rotary methods, with hydraulic rotary currently the most common method. Construction of these types of water wells is covered by American Water Works Association Standards and by an EPA Manual.

Conceptual Design--
Construction costs for wells assume drilling by the hydraulic rotary method in sand and gravel. For small yields (Category 1) a submersible pump is used, while

vertical turbine pumps are assumed for larger flows. Well diameters range from 6 inches at 50 gpm (Category 1), up to 18 to 20 inches for wells in the 500 gpm to 700 gpm category. Construction costs for all well sizes include casing, a well screen, local piping, valves, and electrical work. Wells in Categories 3 and 4 are also assumed to include a well house and significant costs for well development.

Operation and Maintenance Requirements--

Electricity requirements are based on continuous operation of the motor, at a pumping head 50 feet less than the well depth. No energy is included for the housing, as it was assumed that heating and ventilation are unnecessary, and that lighting requirements are minimal. Many wells do not operate continuously and in those cases the energy requirements will be reduced according to the actual load factor. Material requirements are based on necessary lubricants and other routine maintenance items, and servicing the pump and motor once in five years. Labor requirements are based on daily visits for inspection and routine maintenance. Labor and materials required to remove and service the pump and motor once every five years are included in the average annual values.

Bottled Water Vending Machines

For small water systems, the temporary or permanent use of purified water vending machines may be more desirable or cost-effective than constructing and operating treatment systems. Each of the machines used for this purpose is a small physical-chemical treatment system, often designed to remove dissolved minerals and organic compounds as well as microbial contaminants and turbidity.

Conceptual Design--

The purified water machines considered in this document are designed to operate at a flow rate of 2 gpm within an optimum pH range of 4 to 8, and at 110 volts ac. Specific processes assumed to be in the machines include granular activated. carbon, reverse osmosis, ion exchange, mechanical filtration, and UV sterilization. All units contain an automatic shutoff system if the UV light is not operating.

Operation and Maintenance Requirements--

Electrical requirements comprise those necessary to run the machine. Labor and material requirements include such items as replacing filter cartridges, and exchanging carbon and ion exchange canisters.

TYPICAL TREATMENT COSTS FOR SURFACE WATER

Estimated costs for process groups that will reliably remove turbidity and microbial contaminants from surface waters are presented on the following pages. The representative costs contained in these tables are based on all previous assumptions and process descriptions in this Section, and on four additional assumptions:

- These process groups are being added to existing facilities.
- Raw water pumping and finished water pumping either exist or are not required; therefore, no costs are included for these items in the following tables.
- Processes common to all groups include sludge handling and disposal, and clearwell storage.
- System supply capacity for process groups up to 1.0 mgd (Categories 1 through 4) is made up of a combination of a treatment design flow and a treated water storage volume. Design flows and storage for these categories are listed below:

Category	System Supply Capacity, mgd	Treatment Design Capacity, mgd	Supplemental Storage, MG
1	0.07	0.026	0.044
2	0.15	0.068	0.082
3	0.34	0.166	0.174
4	0.84	0.500	0.340

The treatment process design flows are based on multiples of 1.25 to 2.0 times the average flow for each category, as explained in Section V. Estimated costs for supplemental storage are shown in Table VI-3.

Filtration

The following process groups, and the costs thereof, are those intended to remove turbidity and microbial contaminants from water by filtration. The number of individual processes in each group varies from a minimum for slow-sand filters to a maximum for conventional treatment. Disinfection processes and costs are not included in the filtration process groups since they are described separately in subsequent pages.

Package Complete Treatment--

Capital costs, operation and maintenance costs, and total costs (¢/1,000 gallons) are shown in Table VI-4 for package complete treatment for surface water plants in Categories 1 through 6. Processes within the package unit itself include flocculation, tube settling, and mixed-media filtration.

Conventional Treatment--

Estimated costs for a process group providing conventional treatment are shown in Table VI-5. Processes specific to this group include chemical feed, rapid mix, and flocculation, sedimentation, and gravity filtration using mixed media. In common with most filtration process groups, conventional treatment also includes backwashing facilities with washwater surge basins.

Conventional Treatment with Automatic Backwashing Filters--

Estimated costs for conventional treatment with automatic backwashing filters, in place of separate backwashing facilities, are shown in Table VI-6.

Direct Filtration--

Three different modes of direct filtration have been developed for cost estimating purposes, and the results of that analysis are presented in Tables IV-7, VI-8, and VI-9. All three process groups use mixed media for filtration. Costs of installing and operating pressure filters are shown in Table VI-7. Both Tables VI-8 and VI-9 show the costs of using gravity filters, but with two different

TABLE VI-3. ESTIMATED COSTS FOR SUPPLEMENTAL STORAGE FOR SMALL SYSTEMS

Category	Supplemental Storage Volume, gal	Capital Cost $1,000	Total Cost ¢/1,000 gal
1	44,000	57	141.1
2	82,000	73	52.4
3	174,000	128	31.0
4	340,000	224	18.0

Note: O&M costs are considered to be negligible compared to those for other processes and facilities. Pumping costs are not included.

sets of auxiliary processes. Costs in Table VI-8 are for gravity filters preceded by rapid mix and flocculation. Costs in Table VI-9 are for gravity filters preceded by contact basins.

Diatomaceous Earth Filtration--

Costs for diatomaceous earth filtration of surface water in plant Categories 1 through 10 are shown in Table VI-10. Pressure diatomaceous earth filters are assumed in these examples.

Slow-Sand Filtration--

Table VI-11 shows estimated costs for two types of sand filter systems. Costs for systems in Categories 1, 2, and 3 are for covered, concrete filters. Costs for systems in Categories 4 through 7 are for uncovered, membrane-lined earthen berm filters. Operation and maintenance costs in the tables include labor and materials, but no energy costs.

TABLE VI-4. ESTIMATED COSTS FOR SURFACE WATER TREATMENT BY COMPLETE TREATMENT PACKAGE PLANTS

Category	Plant Capacity, mgd	Average Flow, mgd	Capital Cost, $1,000	Operation and Maintenance Cost		Total Cost, ¢/1,000 gal
				$1,000/yr	¢/1,000 gal	
1	0.026	0.013	278	12.2	255.2	944.5
2	0.068	0.045	295	15.9	87 5	277.4
3	0.166	0.13	428	42.4	89.2	195.1
4	0.50	0.40	773	75.1	51.4	113.6
5	2.50	1.30	1,770	137	29.0	72.8
6	5.85	3.25	2,952	274	23.1	52.4

Processes include chemical feed (alum, soda ash, and polymer); complete treatment package plant (flocculation, tube settling and mixed media filtration); backwash storage/clearwell basins; and sludge dewatering lagoons. A separate pumping station is used to transmit unthickened sludge to the sludge dewatering lagoons in Categories 5 and 6. Sludge pumping is included in the cost of the package plant in Categories 1-4. Storage is used to compensate for the lower plant capacities, compared to Table VI-2, for Categories 1-4. The cost of the additional storage is included above.

TABLE VI-5. ESTIMATED COSTS FOR SURFACE WATER TREATMENT BY CONVENTIONAL COMPLETE TREATMENT

Category	Plant Capacity, mgd	Average Flow, mgd	Capital Cost, $1,000	Operation and Maintenance Cost		Total Cost, ¢/1,000 gal
				$1,000/yr	¢/1,000 gal	
5	2.50	2.30	2,819	162.7	34.3	104.1
6	5.85	3.25	4,768	274.1	23.1	70.3
7	11.59	6.75	8,254	474.5	19.3	58.6
8	22.86	11.50	15,533	772.8	18.4	61.9
9	30.68	20.00	22,393	1,294	17.7	53.8
10	109.90	55.50	45,255	2,652	13.1	39.3
11	404	205	129,726	8,714	11.6	32.0
12	1,275	650	395,542	27,034	11.4	31.0

Processes include: Chemical feed (alum, polymer, and sodium hydroxide); rapid mix and flocculation; sedimentation; gravity filtration using mixed media; backwashing with washwater surge basins; in-plant pumping, and sludge pumping and clearwell storage. Categories 5 through 7 include sludge dewatering lagoons and liquid sludge hauling. Categories 8 through 12 include gravity sludge thickeners, filter presses for sludge dewatering, and dewatered sludge hauling to landfill disposal. A building for administration, laboratory, and maintenance purposes is included in all categories.

TABLE VI-6. ESTIMATED COSTS FOR SURFACE WATER TREATMENT BY CONVENTIONAL TREATMENT WITH AUTOMATIC BACKWASHING FILTERS

Category	Plant Capacity, mgd	Average Flow, mgd	Capital Cost, $1,000	Operation and Maintenance Cost		Total Cost, ¢/1,000 gal
				$1,000/yr	¢/1,000 gal	
5	2.50	1.30	2,279	149	31.5	87.9
6	5.85	3.25	3,677	259	21.9	58.3
7	11.59	6.75	6,792	453	18.4	50.8
8	22.86	11.50	14,238	745	17.7	57.6
9	39.68	20.00	20,213	1,235	16.9	49.4
10	109.90	55.50	47,920	2,778	13.7	41.5

Processes for all categories include chemical feed (alum and polymer); rapid mix and flocculation; rectangular clarifiers; continuous automatic backwashing filters, in-plant pumping and sludge pumping and below-grade clearwell storage basin. Categories 5 and 6 include sodium hydroxide feed and sludge dewatering lagoons. Category 7 includes lime feed instead of NaOH, plus liquid sludge hauling. Categories 8 through 10 include gravity sludge thickeners, filter presses for sludge dewatering, and dewatered sludge hauling to landfill disposal. A separate building for administration, laboratory, and maintenance purposes is included in all categories.

TABLE VI-7. ESTIMATED COSTS FOR SURFACE WATER TREATMENT BY DIRECT FILTRATION USING PRESSURE FILTERS

Category	Plant Capacity, mgd	Average Flow, mgd	Capital Cost, $1,000	Operation and Maintenance Cost		Total Cost, ¢/1,000 gal
				$1,000/yr	¢/1,000 gal	
3	0.17	0.13	896	47.9	100.9	322.7
4	0.50	0.40	1,187	60.8	41.7	137.2
5	2.50	1.30	2,080	131	27.6	79.1
6	5.85	3.25	3,118	213	17.9	48.8
7	11.59	6.75	5,084	369	15.0	39.2
8	22.86	11.50	11,386	586	14.0	45.8
9	39.68	20.00	15,684	853	11.7	36.9
10	109.90	55.50	34,235	1,851	8.4	28.2

Processes include chemical feed (alum, polymer, and sodium hydroxide); horizontal pressure filters; mixed media; backwashing and surface wash facilities; washwater surge basins; below-grade clearwell, and sludge pumping. Categories 3 through 7 include sludge dewatering lagoons with sludge hauling. Categories 8 through 10 include gravity sludge thickeners, filter presses, and dewatered sludge hauling to landfill disposal. Storage is used to compensate for the lower plant capacities in Categories 3 and 4, compared to Table VI-2. The cost of the additional storage is included above. A separate building is included in Categories 5-10 for administration, laboratory, and maintenance purposes.

TABLE VI-8. ESTIMATED COSTS FOR SURFACE WATER TREATMENT BY DIRECT FILTRATION USING GRAVITY FILTERS PRECEDED BY FLOCCULATION

Category	Plant Capacity, mgd	Average Flow, mgd	Capital Cost, $1,000	Operation and Maintenance Cost		Total Cost, ¢/1,000 gal
				$1,000/yr	¢/1,000 gal	
4	0.50	0.40	1,266	70.6	48.4	150.2
5	2.50	1.30	2,440	143	30.1	90.5
6	5.85	3.25	3,855	240	20.2	58.4
7	11.59	6.75	6,190	425	17.2	46.8
8	22.86	11.50	12,244	680	16.2	50.5
9	39.68	20.00	16,142	1,014	13.9	39.8
10	109.90	55.50	31,105	2,289	10.6	28.6
11	404	205	89,368	7,151	9.6	23.6
12	1,275	650	242,751	22,010	9.3	21.3

Processes include chemical feed (alum, polymer, and sodium hydroxide); 1-minute rapid mix basin; mechanical flocculation with 30-minute detention, gravity mixed media filters, backwash and surface wash facilities; washwater surge basin; in-plant pumping and sludge pumping, and below-grade clearwell storage basins. Sludge dewatering lagoons and sludge hauling are used in Categories 4 through 7. Categories 8 through 12 use gravity sludge thickeners, filter presses, and dewatered sludge hauling to landfill disposal. Storage is used to compensate for the lower plant capacity in Category 4, compared to Table VI-2. The cost of the additional storage is included above. A separate building for administration, laboratory, and maintenance is included in plants in Categories 5 through 12.

TABLE VI-9. ESTIMATED COSTS FOR SURFACE WATER TREATMENT BY DIRECT FILTRATION USING GRAVITY FILTERS AND CONTACT BASINS

Category	Plant Capacity, mgd	Average Flow, mgd	Capital Cost, $1,000	Operation and Maintenance Cost		Total Cost, ¢/1,000 gal
				$1,000/yr	¢/1,000 gal	
4	0.50	0.40	1,162	55.1	37.7	131.2
5	2.50	1.30	2,242	121	25.4	80.9
6	5.85	3.25	3,714	212	17.9	54.7
7	11.59	6.75	6,047	378	15.4	44.2
8	22.86	11.50	12,060	597	14.2	48.0
9	39.68	20.00	15,867	875	12.0	37.5
10	109.90	55.50	30,255	1,922	8.8	26.3
11	404	205	87,025	5,835	7.8	21.4
12	1,275	650	232,970	17,897	7.6	19.1

Processes include chemical feed (alum, polymer, and sodium hydroxide); 30-minute detention contact basin; gravity filters using mixed media; backwash and surface wash facilities; wash water surge basin; in-plant pumping and sludge pumping; and below-grade clearwell storage basin. Categories 4 through 7 include sludge dewatering lagoons and sludge hauling. Categories 8 through 12 include gravity sludge thickeners, filter presses for sludge dewatering, and dewatered sludge hauling to landfill disposal. Storage is used to compensate for the lower plant capacity for Category 4, compared to Table VI-2. The cost of additional storage is included above. A separate building for administration, laboratory, and maintenance purposes is included for plants in Categories 5-12.

TABLE VI-10. ESTIMATED COSTS FOR SURFACE WATER TREATMENT BY DIRECT FILTRATION USING DIATOMACEOUS EARTH

Category	Plant Capacity, mgd	Average Flow, mgd	Capital Cost, $1,000	Operation and Maintenance Cost		Total Cost, ¢/1,000 gal
				$1,000/yr	¢/1,000 gal	
1	0.026	0.013	221	6.0	127.0	672.9
2	0.068	0.045	285	8.0	43.7	227.2
3	0.166	0.130	374	20.0	42.2	134.7
4	0.50	0.40	570	30.4	20.8	66.6
5	2.50	1.30	1,573	128	27.0	66.0
6	5.85	3.25	2,538	214	18.0	43.1
7	11.59	6.75	4,433	369	15.0	36.1
8	22.86	11.50	10,713	762	18.1	48.1
9	39.68	20.00	15,982	1,165	16.0	41.7
10	109.90	55.50	37,733	2,730	13.5	35.4

Processes include pressure diatomaceous earth filtration units, diatomaceous earth feed equipment; filtered water storage clearwell; and sludge dewatering lagoons. A separate administration, lab, and maintenance building is included in Categories 5-10. Sludge pumps are included in the package facilities used in Categories 1-4, but separate sludge pumping stations are included in Categories 5-10. Categories 8 through 10 include sludge holding tanks, sludge dewatering with filter presses and hauling of dewatered solids to landfill disposal. Storage is used to compensate for the lower plant capacities for Categories 1-4, compared to Table VI-2. The cost of additional storage is included above.

TABLE VI-11. ESTIMATED COSTS FOR SURFACE WATER TREATMENT BY SLOW-SAND FILTRATION

Category	Plant Capacity, mgd	Average Flow, mgd	Capital Cost, $1,000	Operation and Maintenance Cost		Total Cost, ¢/1,000 gal
				$1,000/yr	¢/1,000 gal	
1	0.026	0.013	145	0.8	17.9	377.8
2	0.068	0.045	273	1.6	9.9	205.1
3	0.166	0.133	508	5.1	10.5	133.4
4	0.50	0.40	603	9.0	6.2	54.7
5	2.50	1.30	1,213	20.5	4.3	34.3
6	5.85	3.25	2,573	38.0	3.2	28.7
7	11.59	6.75	4,782	62.3	2.5	25.3

Processes include slow-sand filters and clearwell storage. Sand filters in Categories 1 through 3 are constructed of concrete and are covered. Sand filters in Categories 4 through 7 are constructed of membrane-lined earthen berms and are uncovered. Storage is used to compensate for the lower plant capacities for Category 1 through 4, compared to Table VI-2. The cost of additional storage is included above.

Package Ultrafiltration--
Estimated costs of installation and operation of a package ultrafiltration unit are shown in Table VI-12. Since the potential use of this type of system is predominantly in small water systems, costs are provided only for Categories 1 through 4.

Disinfection

The cost tables include those individual processes intended to destroy microbial contaminants by disinfection. No pretreatment processes or contact chambers are included in the disinfection cost tables. All assumptions defined in the general basis of costs and in the process descriptions apply for these disinfection processes as well as the preceding filtration processes.

Chlorination--
Disinfection costs by chlorination are shown in Table VI-13. Costs include use of hypochlorination for Categories 1 through 3, chlorine feed and cylinder storage for Categories 4 through 10, and chlorine feed and on-site storage for Categories 11 and 12. A dose of 5.0 mg/L is assumed.

Ozone Disinfection--
Table VI-14 shows the cost of disinfecting surface water with ozone. Costs include ozone generation and feed facilities at an application rate of 1.0 mg/L.

Chlorine Dioxide and Chloramination--
Tables VI-15 and VI-16 present costs for chlorine dioxide generation and feed facilities, and chloramination facilities. The latter facilities include storage and feed for both ammonia and chlorine. Application rates are 3.0 mg/L for chlorine dioxide, and a combination of 3.0 mg/L for chlorine and 1.0 mg/L for ammonia in the chloramination process.

Ultraviolet Light--
Cost for disinfection of surface water using ultraviolet light, for Categories 1 through 4, are shown in Table VI-17.

TABLE VI-12. ESTIMATED COSTS FOR SURFACE WATER TREATMENT BY PACKAGE ULTRAFILTRATION PLANTS

Category	Plant Capacity, mgd	Average Flow, mgd	Capital Cost, $1,000	Operation and Maintenance Cost		Total Cost, ¢/1,000 gal
				$1,000/yr	¢/1,000 gal	
1	0.026	0.013	142	5.0	105.2	455.6
2	0.068	0.045	269	9.8	53.7	226.8
3	0.166	0.130	503	26.0	54.7	179.2
4	0.50	0.40	1,144	67.7	46.4	138.4

Processes include a complete package ultrafiltration unit, clearwell storage, and sludge dewatering lagoons with liquid sludge hauling. Storage is used to compensate for the lower plant capacities compared to Table VI-2. The cost of additional storage is included above.

TABLE VI-13. ESTIMATED COSTS FOR SURFACE WATER DISINFECTION USING CHLORINE

Category	Plant Capacity, mgd	Average Flow, mgd	Capital Cost, $1,000	Operation and Maintenance Cost		Total Cost, ¢/1,000 gal
				$1,000/yr	¢/1,000 gal	
1	0.026	0.013	6.1	2.4	50.8	65.9
2	0.068	0.045	12.0	2.5	15.0	23.6
3	0.166	0.133	17.9	5.6	11.9	16.2
4	0.50	0.40	30.2	10.6	7.2	9.7
5	2.50	1.30	45.4	15.2	3.2	4.3
6	5.85	3.25	70.7	24.9	2.1	2.8
7	11.59	6.75	114	39.4	1.6	2.1
8	22.86	11.50	143	50.4	1.2	1.6
9	39.68	20.00	186	73.0	1.0	1.3
10	109.90	55.50	345	162	0.8	1.0
11	404	205	825	524	0.7	0.8
12	1,275	650	1,353	1,424	0.6	0.7

Costs include storage and feed facilities for an application rate of 5.0 mg/L.

TABLE VI-14. ESTIMATED COSTS FOR SURFACE WATER DISINFECTION USING OZONE

Category	Plant Capacity, mgd	Average Flow, mgd	Capital Cost, $1,000	Operation and Maintenance Cost		Total Cost, ¢/1,000 gal
				$1,000/yr	¢/1,000 gal	
1	0.026	0.013	10.0	4.0	84.3	109
2	0.068	0.045	18.5	4.6	25.3	37.2
3	0.166	0.133	32.6	9.2	19.5	27.5
4	0.50	0.40	65.8	10.9	7.4	12.7
5	2.50	1.30	161	14.2	3.0	7.0
6	5.85	3.25	281	20.2	1.7	4.5
7	11.59	6.75	488	27.1	1.1	3.4
8	22.86	11.50	643	33.6	0.8	2.6
9	39.68	20.00	932	51.1	0.7	2.2
10	109.90	55.50	1,897	122	0.6	1.7
11	404	205	4,900	448	0.6	1.4
12	1,275	650	14,141	1,186	0.5	1.2

Costs include ozone generation and feed facilities for an application rate of 1.0 mg/L.

TABLE VI-15. ESTIMATED COSTS FOR SURFACE WATER DISINFECTION USING CHLORINE DIOXIDE

Category	Plant Capacity, mgd	Average Flow, mgd	Capital Cost, $1,000	Operation and Maintenance Cost		Total Cost, ¢/1,000 gal
				$1,000/yr	¢/1,000 gal	
1	0.026	0.013	76.3	6.0	132.8	332
2	0.068	0.045	82.1	6.3	34.8	87.7
3	0.166	0.133	87.3	11.6	24.3	46.1
4	0.50	0.40	102.9	12.2	8.4	16.8
5	2.50	1.30	137	17.1	3.6	7.0
6	5.85	3.25	212	24.9	2.1	4.2
7	11.59	6.75	336	32.0	1.3	2.9
8	22.86	11.50	429	42.0	1.0	2.2
9	39.68	20.00	559	58.4	0.8	1.7
10	109.90	55.50	1,100	142	0.7	1.3
11	404	205	2,210	449	0.6	1.0
12	1,275	650	6,060	1,420	0.6	0.9

Costs include chlorine dioxide generation and feed facilities for an application rate of 3.0 mg/L.

TABLE VI-16. ESTIMATED COSTS FOR SURFACE WATER DISINFECTION BY CHLORAMINATION

Category	Plant Capacity, mgd	Average Flow, mgd	Capital Cost, $1,000	Operation and Maintenance Cost		Total Cost, ¢/1,000 gal
				$1,000/yr	¢/1,000 gal	
1	0.026	0.013	30.2	4.2	87.9	163
2	0.068	0.045	31.8	4.7	28.4	51.1
3	0.166	0.133	34.4	7.4	15.5	23.9
4	0.50	0.40	40.5	16.2	11.1	14.4
5	2.50	1.30	61.6	21.5	4.5	6.1
6	5.85	3.25	101	30.8	2.6	3.6
7	11.59	6.75	147	46.4	1.9	2.6
8	22.86	11.50	214	63.0	1.5	2.1
9	39.68	20.00	289	85.2	1.2	1.6
10	109.90	55.50	557	190	0.9	1.3
11	404	205	875	651	0.9	1.0
12	1,275	650	1,492	2,016	0.8	0.9

Ammonia applied at 1.0 mg/L and chlorine at 3.0 mg/L. Anhydrous ammonia feed facilities are used for Categories 1 through 7, and aqua ammonia facilities for Categories 8 through 12. Chlorine is stored in cylinders in Categories 3 through 9, and in on-site storage tanks in Categories 10 through 12.

TABLE VI-17. ESTIMATED COSTS FOR SURFACE WATER DISINFECTION BY ULTRAVIOLET LIGHT

Category	Plant Capacity, mgd	Average Flow, mgd	Capital Cost, $1,000	Operation and Maintenance Cost		Total Cost, ¢/1,000 gal
				$1,000/yr*	¢/1,000 gal	
1	0.026	0.013	9.6	0.9	25.8	43.2
2	0.068	0.045	11.9	1.2	6.4	14.1
3	0.166	0.133	16.7	2.0	4.1	8.4
4	0.50	0.40	34.0	4.0	2.8	5.4

SUPPLEMENTAL PROCESS COSTS FOR SURFACE WATER TREATMENT

In addition to the complete process groups for which costs have been described above, estimated construction and operation costs have also been developed for a number of individual processes, small process groups, and instrumentation units that could be used to modify or supplement existing surface water treatment processes. Costs for these supplemental processes include no auxiliary facilities except those described below.

Additional Chemical Feed Facilities

Tables VI-18 and VI-19 include estimated costs for adding polymer feed facilities at dosage rates of 0.3 mg/L and 0.5 mg/L, respectively. The tables include facilities, as described earlier in this section, for Categories 1 through 12.

Tables VI-20, VI-21, and VI-22 provide similar cost data for supplemental alum feed, sodium hydroxide feed, and sulfuric acid feed facilities. The assumed dosage rates for alum, sodium hydroxide, and sulfuric acid are 10 mg/L, 10 mg/L, and 2.5 mg/L, respectively.

Modifications to Rapid-Sand Filters

Estimated installation and operation and maintenance costs for two methods of allowing increased application rates to rapid-sand filters are shown in Tables VI-23 and VI-24. Table VI-23 presents costs for capping sand filters with anthracite coal for Categories 1 through 10. Table VI-24 shows analogous installation costs for converting a rapid-sand filter to mixed media.

Adding Tube Settling Modules

Modification of sedimentation basins by adding tube settling modules was described earlier in this section. Estimated installation costs for plant Categories 1 through 12 are shown in Table VI-25.

TABLE VI-18. ESTIMATED COSTS FOR SURFACE WATER TREATMENT BY ADDING POLYMER FEED FACILITIES (0.3 mg/L)

Category	Plant Capacity, mgd	Average Flow, mgd	Capital Cost, $1,000	Operation and Maintenance Cost $1,000/yr*	Operation and Maintenance Cost ¢/1,000 gal	Total Cost, ¢/1,000 gal
1	0.026	0.013	2.8	1.3	28.3	35.3
2	0.068	0.045	3.3	1.4	8.6	11.0
3	0.166	0.13	3.8	3.5	7.3	8.2
4	0.50	0.40	4.6	3.7	2.5	2.9
5	2.50	1.30	66.1	5.8	1.2	2.9
6	5.85	3.25	66.7	6.8	0.6	1.2
7	11.59	6.75	68.8	8.4	0.3	0.7
8	22.86	11.50	72.7	10.7	0.2	0.5
9	39.68	20.00	78.6	14.6	0.2	0.3
10	109.90	55.50	107.2	31.1	0.2	0.2
11	404	205	159.9	100	0.1	0.2
12	1,275	650	227.3	299	0.1	0.1

*Polymer cost for Categories 1 through 4 is 75¢/lb, and for Categories 5 through 12 it is $1,000/ton.

TABLE VI-19. ESTIMATED COSTS FOR SURFACE WATER TREATMENT BY ADDING POLYMER FEED FACILITIES (0.5 mg/L)

Category	Plant Capacity, mgd	Average Flow, mgd	Capital Cost, $1,000	Operation and Maintenance Cost		Total Cost, ¢/1,000 gal
				$1,000/yr*	¢/1,000 gal	
1	0.026	0.013	3.0	1.4	28.9	36.4
2	0.068	0.045	3.6	1.5	8.8	11.4
3	0.166	0.13	4.1	3.6	7.4	8.4
4	0.50	0.40	5.2	3.8	2.6	3.0
5	2.50	1.30	66.1	6.2	1.3	3.0
6	5.85	3.25	68.2	7.8	0.7	1.4
7	11.59	6.75	71.6	10.6	0.4	0.8
8	22.86	11.50	78.2	14.2	0.3	0.6
9	39.68	20.00	90.9	20.8	0.3	0.4
10	109.90	55.50	125.4	48.7	0.2	0.3
11	404	205	186.8	164	0.2	0.2
12	1,275	650	265.9	506	0.2	0.2

*Polymer cost for Categories 1 through 4 is 75¢/lb, and for Categories 5 through 12 it is $1,000/ton.

TABLE VI-20. ESTIMATED COSTS FOR SURFACE WATER TREATMENT BY ADDING ALUM FEED FACILITIES (10 MG/L)

Category	Plant Capacity, mgd	Average Flow, mgd	Capital Cost, $1,000	Operation and Maintenance Cost		Total Cost, ¢/1,000 gal
				$1,000/yr	¢/1,000 gal	
1	0.026	0.13	12.2	2.3	47.8	87.1
2	0.068	0.45	15.9	2.6	16.0	26.2
3	0.166	0.13	20.3	5.4	11.5	16.5
4	0.50	0.40	28.0	7.7	5.3	7.5
5	2.50	1.30	40.7	4.1	0.9	1.9
6	5.85	3.25	47.1	7.9	0.7	1.1
7	11.59	6.75	58.1	14.9	0.6	0.9
8	22.86	11.50	70.4	24.4	0.6	0.8
9	39.68	20.00	81.6	41.3	0.6	0.7
10	109.90	55.50	128	112	0.55	0.63
11	404	205	250	402	0.54	0.58
12	1,275	650	575	1,245	0.52	0.55

TABLE VI-21. ESTIMATED COSTS FOR SUPPLEMENTING SURFACE WATER TREATMENT BY ADDING SODIUM HYDROXIDE FEED FACILITIES

Category	Plant Capacity, mgd	Average Flow, mgd	Capital Cost, $1,000	Operation and Maintenance Cost		Total Cost, ¢/1,000 gal
				$1,000/yr	¢/1,000 gal	
1	0.026	0.013	2.4	1.2	24.3	30.2
2	0.068	0.045	2.7	1.4	8.8	10.7
3	0.166	0.13	3.3	3.3	7.6	8.4
4	0.50	0.40	5.4	6.2	4.2	4.7
5	2.50	1.30	33.3	8.8	1.8	2.7
6	5.85	3.25	36.9	18.4	1.6	1.9
7	11.59	6.75	43.0	35.8	1.4	1.6
8	22.86	11.50	56.3	59.4	1.4	1.6
9	39.68	20.00	76.4	101	1.4	1.5
10	109.90	55.50	159	275	1.4	1.4
11	404	205	353	1,015	1.4	1.4
12	1,275	650	697	3,210	1.4	1.4

Costs include storage and feed facilities to add NaOH at a concentration of 10 mg/L. Dry sodium hydroxide is used for Categories 1 through 4, while a liquid solution is used for bulk delivery for Categories 5 through 12.

TABLE VI-22. ESTIMATED COSTS FOR SUPPLEMENTING SURFACE WATER TREATMENT BY ADDING SULFURIC ACID FEED FACILITIES

Category	Plant Capacity, mgd	Average Flow, mgd	Capital Cost, $1,000	Operation and Maintenance Cost		Total Cost, ¢/1,000 gal
				$1,000/yr	¢/1,000 gal	
1	0.026	0.013	2.8	0.9	20.0	27.0
2	0.068	0.045	3.2	1.3	7.6	9.9
3	0.166	0.13	4.0	3.2	6.7	7.6
4	0.50	0.40	5.7	5.6	3.9	4.3
5	2.50	1.30	21.6	2.4	0.5	1.0
6	5.85	3.25	25.6	3.6	0.3	0.6
7	11.59	6.75	29.7	5.8	0.2	0.4
8	22.86	11.50	37.9	8.8	0.2	0.3
9	39.68	20.00	49.9	14.1	0.2	0.3
10	109.90	55.50	86.9	36.3	0.2	0.2
11	404	205	210.3	128.3	0.2	0.2
12	1,275	650	431.1	404.7	0.2	0.2

Costs include storage (15 days), feed, and metering facilities for delivering concentrated acid directly from storage to application point. Categories 1 through 6 include delivery in drums, stored indoors. Categories 7 through 12 include bulk delivery and outdoor storage in FRP tanks. Application rate is 2.5 mg/L.

TABLE VI-23. ESTIMATED COSTS FOR UPGRADING SURFACE WATER TREATMENT BY CAPPING RAPID-SAND FILTERS WITH ANTHRACITE COAL

Category	Plant Capacity, mgd	Average Flow, mgd	Capital Cost, $1,000	Operation and Maintenance Cost		Total Cost, ¢/1,000 gal
				$1,000/yr	¢/1,000 gal	
1	0.026	0.013	0.2	0	0	0.5
2	0.068	0.045	0.4	0	0	0.3
3	0.166	0.133	1.1	0	0	0.3
4	0.50	0.40	3.1	0	0	0.3
5	2.50	1.30	12.4	0	0	0.3
6	5.85	3.25	27.4	0	0	0.3
7	11.59	6.75	52.0	0	0	0.3
8	22.86	11.50	102.5	0	0	0.3
9	39.68	20.00	178.1	0	0	0.3
10	109.90	55.50	485.7	0	0	0.3

Costs include sand removal and on-site disposal, material and freight costs for anthracite coal, and installation labor.

TABLE VI-24. ESTIMATED COSTS FOR UPGRADING SURFACE WATER TREATMENT BY CONVERTING RAPID-SAND FILTERS TO MIXED MEDIA FILTERS

Category	Plant Capacity, mgd	Average Flow, mgd	Capital Cost, $1,000	Operation and Maintenance Cost		Total Cost, ¢/1,000 gal
				$1,000/yr	¢/1,000 gal	
1	0.026	0.013	3.8	0	0	9.4
2	0.068	0.045	7.7	0	0	5.5
3	0.166	0.133	13.5	0	0	3.3
4	0.50	0.40	26.1	0	0	2.1
5	2.50	1.30	82.8	0	0	2.0
6	5.85	3.25	169.1	0	0	1.7
7	11.59	6.75	335.6	0	0	1.6
8	22.86	11.50	546.7	0	0	1.5
9	39.68	20.00	942.8	0	0	1.5
10	109.90	55.50	2,584	0	0	1.5

Costs include removal and disposal of sand and gravel, and installation of new media.

TABLE VI-25. ESTIMATED COSTS FOR UPGRADING SURFACE WATER TREATMENT BY ADDING TUBE SETTLING MODULES

Category	Plant Capacity, mgd	Average Flow, mgd	Capital Cost, $1,000	Operation and Maintenance Cost		Total Cost, ¢/1,000 gal
				$1,000/Yr	¢/1,000 Gal	
1	0.026	0.013	1.1	0	-	2.7
2	0.068	0.045	2.3	0	-	1.6
3	0.166	0.133	3.9	0	-	0.9
4	0.50	0.40	8.2	0	-	0.7
5	2.50	1.30	25.2	0	-	0.6
6	5.85	3.25	46.8	0	-	0.5
7	11.59	6.75	92.0	0	-	0.4
8	22.86	11.50	163.0	0	-	0.4
9	39.68	20.00	250	0	-	0.4
10	109.90	55.50	684	0	-	0.4
11	404	205	2,500	0	-	0.4
12	1,275	650	7,850	0	-	0.4

Costs include modules, supports and anchor brackets, transition baffle, effluent launders, and installation.

Direct Filtration Modifications

Tables VI-26 through VI-29 provide estimated costs for potential methods of adding processes upstream of filters in a direct filtration plant. Table VI-26 shows costs for installing a contact basin. Tables VI-27 through VI-29 show costs for adding rapid-mix, flocculation, and clarification facilities, respectively. The capital costs in all tables include costs for site work involving connections with existing piping.

Additional Filtration Facilities

Estimated costs for two additional sets of facilities that are optional auxiliary components for filtration are shown in Tables VI-30 and VI-31. Costs for hydraulic surface wash facilities are presented in Table VI-30 and include pumps, electrical control, piping, valves, and headers within the filter pipe gallery. Costs for filter-to-waste facilities (Table VI-31) include all necessary valves, controls, and piping within the filter gallery, but do not include piping outside the filter area.

Finished Water Pumping

As noted earlier in this section, none of the filtration process group costs or the disinfection process costs include any external pumping, since it is assumed that these processes and process groups are being added to existing facilities that have the necessary pumping. In case finished water pumping does need to be added, however, Table VI-32 presents the costs for constructing and operating such a facility. Costs for Categories 1 through 4 assume use of a package high-service pumping station, while Categories 5 through 12 assume use of a custom-designed station.

Additional Instrumentation

The installation costs for two types of instruments capable of providing significant improvement in plant performance are shown in Tables VI-33 and VI-34. Table

TABLE VI-26. ESTIMATED COSTS FOR UPGRADING SURFACE WATER TREATMENT BY ADDING CONTACT BASINS TO A DIRECT FILTRATION PLANT

Category	Plant Capacity, mgd	Average Flow, mgd	Capital Cost, $1,000	Operation and Maintenance Cost		Total Cost, ¢/1,000 gal
				$1,000/yr	¢/1,000 gal	
1	0.026	0.013	6.5	0	0	16.1
2	0.068	0.045	15.2	0	0	10.9
3	0.166	0.133	24.3	0	0	6.0
4	0.50	0.40	44.3	0	0	3.6
5	2.50	1.30	145.0	0	0	3.68
6	5.85	3.25	259.6	0	0	2.61
7	11.59	6.75	371.8	0	0	1.80
8	22.86	11.50	551.9	0	0	1.59
9	39.68	20.00	796.1	0	0	1.30
10	109.90	55.50	1,592	0	0	0.96
11	404	205	5,799	0	0	0.96
12	1,275	650	11,270	0	0	0.96

Costs are based on contact basin with 30-minute detention time.

TABLE VI-27. ESTIMATED COSTS FOR SUPPLEMENTING SURFACE WATER TREATMENT BY ADDING RAPID MIX

Category	Plant Capacity, mgd	Average Flow, mgd	Capital Cost, $1,000	Operation and Maintenance Cost $1,000/yr	Operation and Maintenance Cost ¢/1,000 gal	Total Cost, ¢/1,000 gal
1	0.026	0.013	13.2	2.8	58.6	91.3
2	0.068	0.045	17.5	2.9	17.6	30.1
3	0.166	0.133	22.5	7.0	14.7	20.3
4	0.50	0.40	30.9	7.9	5.4	7.9
5	2.50	1.30	47.7	13.3	2.8	4.0
6	5.85	3.25	63.7	22.4	1.9	2.5
7	11.59	6.75	88.2	38.2	1.6	2.0
8	22.86	11.50	139	69.2	1.6	2.0
9	39.68	20.00	218	116	1.6	1.9
10	109.90	55.50	587	313	1.5	1.9
11	404	205	2,100	1,130	1.5	1.8
12	1,275	650	6,670	3,540	1.5	1.8

TABLE VI-28. ESTIMATED COSTS FOR SUPPLEMENTING SURFACE WATER TREATMENT BY ADDING FLOCCULATION

Category	Plant Capacity, mgd	Average Flow, mgd	Capital Cost, $1,000	Operation and Maintenance Cost		Total Cost, ¢/1,000 gal
				$1,000/yr	¢/1,000 gal	
1	0.026	0.013	10	1.0	21.7	45.2
2	0.068	0.045	18	1.1	6.9	20.1
3	0.166	0.133	34	2.3	4.9	13.3
4	0.50	0.40	73	2.7	1.8	7.7
5	2.50	1.30	217	3.8	0.8	6.2
6	5.85	3.25	325	5.6	0.5	3.7
7	11.59	6.75	418	8.7	0.4	2.3
8	22.86	11.50	587	14.5	0.4	2.0
9	39.68	20.00	840	22.9	0.3	1.7
10	109.90	55.50	1,830	53.9	0.3	1.3
11	404	205	6,060	182	0.2	1.2
12	1,275	650	19,200	569	0.2	1.2

TABLE VI-29. ESTIMATED COSTS FOR SUPPLEMENTING SURFACE WATER TREATMENT BY ADDING RECTANGULAR CLARIFIERS

Category	Plant Capacity, mgd	Average Flow, mgd	Capital Cost, $1,000	Operation and Maintenance Cost		Total Cost, ¢/1,000 gal
				$1,000/yr	¢/1,000 gal	
1	0.026	0.013	28	1.2	25.9	95.5
2	0.068	0.045	46	1.4	8.7	41.7
3	0.166	0.133	102	5.3	11.2	36.4
4	0.50	0.40	174	6.3	4.3	18.3
5	2.50	1.30	435	9.2	2.0	12.7
6	5.85	3.25	960	19.9	1.7	11.2
7	11.59	6.75	1,930	37.0	1.5	10.7
8	22.86	11.50	3,110	58.8	1.4	10.1
9	39.68	20.00	5,220	102	1.4	9.8
10	109.90	55.50	14,300	244	1.2	9.5
11	404	205	51,000	858	1.2	9.2
12	1,275	650	162,000	2,720	1.2	9.2

TABLE VI-30. ESTIMATED COSTS FOR ADDING HYDRAULIC SURFACE WASH FACILITIES

Category	Plant Capacity, mgd	Average Flow, mgd	Capital Cost, $1,000	Operation and Maintenance Cost		Total Cost, ¢/1,000 gal
				$1,000/yr	¢/1,000 gal	
1	0.026	0.013	27.6	0.6	11.8	80.1
2	0.068	0.045	35.3	0.6	3.8	29.0
3	0.166	0.133	43.5	1.2	2.5	13.3
4	0.50	0.40	56.4	1.4	0.99	5.5
5	2.50	1.30	80.9	2.1	0.45	2.4
6	5.85	3.25	114	3.2	0.27	1.4
7	11.59	6.75	187	4.8	0.20	1.1
8	22.86	11.50	247	7.7	0.18	0.9
9	39.68	20.00	360	11.7	0.16	0.7
10	109.90	55.50	950	28.4	0.14	0.7
11	404	205	3,310	101	0.14	0.7
12	1,275	650	10,100	343	0.14	0.6

TABLE VI-31. ESTIMATED COSTS FOR ADDING FILTER-TO-WASTE FACILITIES

Category	Plant Capacity, mgd	Average Flow, mgd	Capital Cost, $1,000	Operation and Maintenance Cost		Total Cost, ¢/1,000 gal
				$1,000/yr	¢/1,000 gal	
1	0.03	0.01	2.2	0	0	5.5
2	0.07	0.05	5.6	0	0	4.0
3	0.17	0.13	9.9	0	0	2.4
4	0.50	0.40	16.5	0	0	1.3
5	2.50	1.30	36.4	0	0	0.9
6	5.85	3.25	40.4	0	0	0.4
7	11.59	6.75	54.5	0	0	0.3
8	22.86	11.50	71.5	0	0	0.2
9	39.68	20.00	124	0	0	0.2
10	109.90	55.50	345	0	0	0.2
11	404	205	759	0	0	0.1
12	1,275	650	2,410	0	0	0.1

Note: Assumes one backwash unit for Category 1, two for Categories 2-4, four for Categories 5-8, six for Category 9, 16 for Category 10, 56 for Category 11, and 178 for Category 12. Costs include valves and controls, plus 10 ft of pipe and fittings.

TABLE VI-32. ESTIMATED COSTS FOR FINISHED WATER PUMPING

Category	Plant Capacity, mgd	Average Flow, mgd	Capital Cost, $1,000*	Operation and Maintenance Cost		Total Cost, ¢/1,000 gal
				$1,000/yr	¢/1,000 gal	
1	0.03	0.01	22.6	0.7	14.3	70.3
2	0.07	0.05	24.6	1.0	6.1	23.7
3	0.17	0.13	27.0	2.9	6.0	12.7
4	0.50	0.40	33.6	6.3	4.3	7.0
5	2.50	1.30	243	58.5	12.3	18.4
6	5.85	3.25	457	135	11.4	16.0
7	11.59	6.75	792	272	11.1	14.9
8	22.86	11.50	1,215	459	10.9	14.3
9	39.68	20.00	1,615	792	10.9	13.9
10	109.90	55.50	5,055	2,184	10.8	13.7
11	404	205	18,150	8,048	10.8	13.7
12	1,275	650	56,340	25,508	10.8	13.7

*Assumes factor of 1.48 times construction cost for sitework, contractors overhead and profit, engineering, legal and administrative costs, and interest during construction.

Note: Categories 1 through 4 assume use of package high service pump station (maximum output pressure = 70 psi). Categories 5 through 12 assume use of custom-built finished water pumping station operating at 300 ft TDH. Costs for other head conditions are included in Appendix E.

TABLE VI-33. ESTIMATED COSTS FOR IMPROVING SURFACE WATER TREATMENT BY ADDING FILTER EFFLUENT TURBIDIMETERS

Category	Plant Capacity, mgd	Average Flow, mgd	Capital Cost, $1,000	Operation and Maintenance Cost		Total Cost, ¢/1,000 gal
				$1,000/yr	¢/1,000 gal	
4	0.50	0.40	27.0	0	-	2.2
5	2.50	1.30	45.0	0	-	1.1
6	5.85	3.25	45.0	0	-	0.4
7	11.59	6.75	81.0	0	-	0.4
8	22.86	11.50	81.0	0	-	0.2
9	39.68	20.00	81.0	0	-	0.1
10	109.90	55.50	135	0	-	0.08
11	404	205	225	0	-	0.04
12	1,275	650	405	0	-	0.02

Costs include turbidimeters, sampling pumps, piping, and instrumentation. One system assumed installation in clearwell, plus on the effluent lines of two filters in Category 4, four filters in Categories 5 and 6, eight filters in Categories 7 through 9, 14 filters in Category 10, 24 filters in Category 11, and 44 filters in Category 12. A single complete system would have a capital cost of approximately $9,000.

TABLE VI-34. ESTIMATED COSTS FOR COAGULANT CONTROL SYSTEM

Cost Category	Cost, $
Manufactured equipment*	$ 9,000
Electrical and instrumentation	1,000
Installation	2,500
Subtotal	$12,500
Miscellaneous and Contingency	1,900
Construction Cost	$14,400
Engineering, legal, financial, administrative	4,600
TOTAL CAPITAL COST	$19,000

*Includes coagulant control system, 2-pen recorder, and flow controller. Assumes metering pump already in place. Normal maintenance should average less than $100 per year for entire system.

VI-33 presents costs for filter effluent turbidimeters, while Table VI-34 tabulates the installation cost of a single coagulant control system. This system can be used to monitor the effectiveness of flocculant/coagulant dosages; thereby enabling plant operators to adjust chemical dosage rates to obtain optimum performance.

Alternatives to Treatment

In some cases, it may be more cost-effective for a small water system to choose an alternative course of action, rather than constructing a treatment system. Two examples of such alternatives are either to construct a new well or to purchase and install bottled water vending machines. Costs for designing, constructing, and operating a well 350 feet deep are presented in Table VI-35. Costs for implementing use of enough bottled water vending machines to supply the design flow for Categories 1 and 2 are presented in Table VI-36. Assumptions used in estimating costs of these alternatives are those presented on previous pages, including the general basis of costs and conceptual design and operating requirements.

COST SUMMARY

A comparison of the total costs of all treatment processes listed in this section is provided in Table VI-37. Values in the table are taken from the last column of the cost table of each treatment process and process group, and are expressed in units of cents per 1,000 gallons.

TABLE VI-35. ESTIMATED COSTS FOR CONSTRUCTING A NEW WELL

Category	Plant Capacity, mgd	Average Flow, mgd	Capital Cost, $1,000	Operation and Maintenance Cost $1,000/yr	Operation and Maintenance Cost ¢/1,000 gal	Total Cost, ¢/1,000 gal
1	0.07	0.01	56	7.8	164.4	303.0
2	0.15	0.05	75	10.5	63.9	117.5
3	0.34	0.13	115	15.4	31.7	59.5
4	0.84	0.40	205	27.0	18.5	35.0

Costs are for a well 350 feet deep, and include engineering costs.

TABLE VI-36. ESTIMATED COSTS FOR PURCHASING AND USING BOTTLED WATER VENDING MACHINES

Category	Plant Capacity, mgd	Average Flow, mgd	Capital Cost, $1,000	Operation and Maintenance Cost		Total Cost, ¢/1,000 gal
				$1,000/yr	¢/1,000 gal	
1	0.07	0.01	215	12.5	263.4	795.5
2	0.15	0.05	447	26.0	158.3	478.1

TABLE VI-37. SUMMARY OF TOTAL COSTS

Treatment Processes	Total Cost of Treatment, ¢/1,000 Gallons Size Category[1]											
	1	2	3	4	5	6	7	8	9	10	11	12
	0.026	0.068	0.166	0.50	2.50	5.85	11.59	22.86	39.68	109.9	404	1,275
	0.013	0.045	0.133	0.40	1.30	3.25	6.75	11.50	20.00	55.5	205	650
Filtration[2]												
Complete treatment package plants	944.5	277.4	195.1	113.6	72.8	52.4						
Conventional complete treatment					104.1	70.3	58.6	61.9	53.8	39.3	32.0	31.0
Conventional treatment with automatic backwashing filters					87.9	58.3	50.8	57.6	49.4	41.5		
Direct filtration using pressure filters			322.7	137.2	79.1	48.8	39.2	45.8	36.9	28.2		
Direct filtration using gravity filters preceded by flocculation				150.2	90.5	58.4	46.8	50.5	39.8	28.6	23.6	21.3
Direct filtration using gravity filters and contact basins				131.2	80.9	54.7	44.2	48.0	37.5	26.3	21.4	19.1
Direct filtration using diatomaceous earth	672.9	227.2	134.7	66.6	43.1	43.1	36.1	48.1	41.7	35.4		
Slow-sand filtration	377.8	205.1	133.4	54.7	34.3	28.7	25.3					
Package ultrafiltration plants	455.6	226.8	179.2	138.4								

1. Category values, from top to bottom, are number, design flow (mgd), and average flow (mgd). Population ranges for each category are:

1. 25 - 100	4. 1,001 - 3,300	7. 25,001 - 50,000	10. 100,001 - 500,000
2. 101 - 500	5. 3,301 - 10,000	8. 50,001 - 75,000	11. 500,001 - 1,000,000
3. 501 - 1,000	6. 10,001 - 25,000	9. 75,001 - 100,000	12. >1,000,000

2. Each process group includes chemical addition and individual liquid and solids handling processes required for operation; excluded are raw water pumping, finished water pumping, and disinfection.

TABLE VI-37 (Continued)

Treatment Processes	Total Cost of Treatment, ¢/1,000 Gallons Size Category[1]											
	1	2	3	4	5	6	7	8	9	10	11	12
	0.026	0.068	0.166	0.50	2.50	5.85	11.59	22.86	39.68	109.9	404	1,275
	0.013	0.045	0.133	0.40	1.30	3.25	6.75	11.50	20.00	55.5	205	650
Disinfection[3]												
Chlorine feed facilities[4]	65.9	23.6	16.2	9.7	4.3	2.8	2.1	1.6	1.3	1.0	0.8	0.7
Ozone generation and feed[5]	109	37.2	27.5	12.7	7.0	4.5	3.4	2.6	2.2	1.7	1.4	1.2
Chlorine dioxide[6]	322	87.7	46.1	16.8	7.0	4.2	2.9	2.2	1.7	1.3	1.0	0.9
Chloramination[7]	163	51.1	23.9	14.4	6.1	3.6	2.6	2.1	1.6	1.3	1.0	0.9
Ultraviolet light	43.2	14.1	8.4	5.4								
Supplemental Processes												
Add polymer feed, 0.3 mg/L	35.3	11.0	8.2	2.9	2.9	1.2	0.7	0.5	0.3	0.2	0.2	0.1
Add polymer feed, 0.5 mg/L	36.4	11.4	8.4	3.0	3.0	1.4	0.8	0.6	0.4	0.3	0.2	0.2
Add alum feed, 10 mg/L	87.1	26.2	16.5	7.5	1.9	1.1	0.9	0.8	0.7	0.6	0.6	0.5
Add sodium hydroxide feed	30.2	10.7	8.4	4.7	2.7	1.9	1.6	1.6	1.5	1.4	1.4	1.4
Add sulfuric acid feed	27.0	9.9	7.6	4.3	1.0	0.6	0.4	0.3	0.3	0.2	0.2	0.2
Capping rapid-sand filters with anthracite coal	0.5	0.3	0.3	0.3	0.3	0.3	0.3	0.3	0.3	0.3		
Converting rapid-sand filters to mixed-media filters	9.4	5.5	3.3	2.1	2.0	1.7	1.6	1.5	1.5	1.5		

3. Disinfection facilities include all required generation, storage, and feed equipment; contact basin and detention facilities are excluded. Design flows for Categories 1-4 are, respectively: 0.026 mgd, 0.068 mgd, 0.166 mgd, and 0.50 mgd.
4. Dose is 5.0 mg/L; includes hypochlorite solution feed for Categories 1-3, chlorine feed and cylinder storage for Categories 4-10, and chlorine feed and on-site storage for Categories 11 and 12.
5. Dose is 1.0 mg/L.
6. Dose is 3.0 mg/L.
7. Doses are chlorine at 3.0 mg/L and ammonia at 1.0 mg/L.

TABLE VI-37 (Continued)

Treatment Processes	Total Cost of Treatment, ¢/1,000 Gallons Size Category[1]											
	1	2	3	4	5	6	7	8	9	10	11	12
	0.026	0.068	0.166	0.50	2.50	5.85	11.59	22.86	39.68	109.9	404	1,275
	0.013	0.045	0.133	0.40	1.30	3.25	6.75	11.50	20.00	55.5	205	650
Supplemental Processes (cont.)												
Add tube settling modules	2.7	1.6	0.9	0.7	0.6	0.5	0.4	0.4	0.4	0.4	0.4	0.4
Add contact basins to an in-line direct filtration plant	16.1	10.9	6.0	3.6	3.7	2.6	1.8	1.6	1.3	1.0	1.0	1.0
Add rapid mix	91.3	30.1	20.3	7.9	4.0	2.5	2.0	2.0	1.9	1.9	1.8	1.8
Add flocculation	45.2	20.1	13.3	7.7	6.2	3.7	2.3	2.0	1.7	1.3	1.2	1.2
Add clarification	95.5	41.7	36.4	18.3	12.7	11.2	10.7	10.1	9.8	9.5	9.2	9.2
Add hydraulic surface wash	80.1	29.0	13.3	5.5	2.4	1.4	1.1	0.9	0.7	0.7	0.7	0.6
Add filter-to-waste facilities	5.5	4.0	2.4	1.3	0.9	0.4	0.3	0.2	0.2	0.2	0.1	0.1
Finished water pumping[8]	70.3	23.7	12.7	7.0	18.4	16.0	14.9	14.3	13.9	13.7	13.7	13.7
Alternatives to Treatment[9]												
Construct new well, 350 ft	303.0	117.5	59.5	35.0								
Bottled water vending machines	795.5	478.1										

8. Facilities include a package high service pumping station for Categories 1-4, and a custom-designed and constructed station for Categories 5-12.
9. Design flows are equal to system demand, i.e.,

Category	Design Flow, mgd	Average Flow, mgd
1	0.07	0.013
2	0.15	0.045
3	0.34	0.133
4	0.84	0.400

References

EXECUTIVE SUMMARY

1. Lippy, E.C., and S.C. Waltrip. "Waterborne Disease Outbreaks--1946-1980: A Thirty-Five Year Perspective." **J.AWWA**, p. 60, February, 1984.

2. Craun, G.F., and Jakubowski, W. "Status of Waterborne Giardiasis Outbreaks and Monitoring Methods." American Water Resources Assoc., Water Related Health Issues Symposium, November, 1986, Atlanta.

3. Logsdon, G.S. "Comparison of Some Filtration Processes Appropriate for Giardia Cyst Removal." Presented at the Calgary Giardia Conference, Calgary, Alberta, Canada, February 23-25, 1987.

4. Amirtharajah, A. "Variance Analyses and Criteria for Treatment Regulations." **J.AWWA**, 78:34-49, March, 1986.

5. DeWalle, F.B., J. Engeset, and W. Lawrence. "Removal of Giardia Lamblia Cysts by Drinking Water Plants." EPA-600/S2-84-069, U.S. Environmental Protection Agency, MERL, Cincinnati, OH, May, 1984.

6. Al-Ani, M., J.M. McElroy, C.P. Hibler, and D.W. Hendricks. "Filtration of Giardia Cysts and Other Substances, Vol. 3: Rapid Rate Filtration." EPA-600/2-85-027, U.S. Environmental Protection Agency, MERL, Cincinnati, OH, April, 1985. NTIS: PB-85-194645/AS.

7. Lange, K.P., W.D. Bellamy, and D.W. Hendricks. "Filtration of Giardia Cysts and Other Substances, Vol. 1: Diatomaceous Earth Filtration." EPA-600/2-84-114, U.S. Environmental Protection Agency, MERL, Cincinnati, OH, June, 1984. NTIS: PB-84-212703.

8. Bellamy, W.D., G.P. Silverman, and D.W. Hendricks. "Filtration of Giardia Cysts and Other Substances, Vol. 2: Slow Sand Filtration." EPA-600/2-85-026, U.S. Environmental Protection Agency, MERL, Cincinnati, OH, April, 1985. NTIS: PB-85-191633/AS.

EXECUTIVE SUMMARY (Continued)

9. Culp, R.L. "Direct Filtration." J.AWWA, p. 357, July, 1977.

10. Hoff, J.C. "Inactivation of Microbial Agents by Chemical Disinfectants." EPA/600/S2-86/067, U.S. Environmental Protection Agency, Water Engineering Research Laboratory, Cincinnati, OH, September, 1986.

11. Malcolm Pirnie, Inc. "Analysis of Plant Capacities and Average Flow Rates." Letter report prepared for the U.S. Environmental Protection Agency, Office of Drinking Water, Washington, D.C., January 6, 1986.

12. Malcolm Pirnie, Inc., and Consumers/Pirnie Utility Services Company. "Evaluation of Costs Associated With Turbidity Removal Systems in Small Water Companies." Draft report prepared for the U.S. Environmental Protection Agency, July, 1986.

13. Culp/Wesner/Culp. "Estimation of Small System Water Treatment Costs." EPA Contract No. 68-03-3093, U.S. Environmental Protection Agency, Cincinnati, OH. In publication.

14. Culp/Wesner/Culp. "Estimating Water Treatment Costs, Volume 2: Cost Curves Applicable to 1 to 200 mgd Treatment Plants." EPA-600/2-79-162b, U.S. Environmental Protection Agency, Cincinnati, OH, August, 1979.

15. Malcolm Pirnie, Inc. "Analysis of Plant Capacities and Average Flow Rates." Letter report prepared for the U.S. Environmental Protection Agency, Office of Drinking Water, Washington, D.C., January 6, 1986.

16. Taylor, E.W. "Forty-fifth Report on the Results of the Bacteriological, Chemical, and Biological Examination of the London Waters for the Years 1971-73." Metropolitan Water Board, London, England, no date.

17. Malina, J.F., Jr., B.D. Moore, and J.L. Marshall. "Poliovirus Removal by Diatomaceous Earth Filtration." Center for Research in Water Resources, The University of Texas, Austin, Texas, 1972.

EXECUTIVE SUMMARY (Continued)

18. Robeck, G.G., N.A. Clarke, and K.A. Dostal. "Effectiveness of Water Treatment Processes in Virus Removal." **J.AWWA,** 54(10):1275-1290, 1962.

SECTION II

1. Lippy, E.C., and S.C. Waltrip. "Waterborne Disease Outbreaks--1946-1980: A Thirty-Five Year Perspective." **J.AWWA,** p. 60, February, 1984.

2. Craun, G.F., and Jakubowski, W. "Status of Waterborne Giardiasis Outbreaks and Monitoring Methods." American Water Resources Assoc., Water Related Health Issues Symposium, November, 1986, Atlanta.

3. Center for Disease Control. "Foodborne and Waterborne Disease Outbreaks, Annual Summaries, 1981, 1982, 1983." U.S. Department of Health, Education and Welfare, Atlanta, GA.

4. Committee on the Status of Waterborne Diseases in the United States and Canada. "Waterborne Disease in the United States and Canada." **J.AWWA,** pp. 528-529, October, 1981.

5. Logsdon, G.S., and E.C. Lippy. "The Role of Filtration in Preventing Waterborne Disease." **J.AWWA,** p. 649, December, 1982.

6. Pluntze, J.C. "The Need for Filtration of Surface Water Supplies, Viewpoint." **J.AWWA,** p. 11, December, 1984.

7. Jakubowski, W., et al. "Waterborne Giardia: It's Enough to Make You Sick, Roundtable." **J.AWWA,** p. 14, February, 1985.

8. "Interim Primary Drinking Water Regulations: Amendments." U.S. Environmental Protection Agency, 45(168):57332-57, 1980.

9. Ward, R.L., and E.W. Akin. "Minimum Infective Dose of Animal Viruses." **CRC Critical Reviews in Environmental Control,** 14(4):297-310, CRC Press, Inc., 1984.

SECTION II (Continued)

10. "National Interim Primary Drinking Water Regulations." U.S. Environmental Protection Agency, EPA-570/9-76-003.

11. Fair, G.M., et al. **Water Supply and Waste Disposal.** John Wiley & Sons, New York, NY, p. 552, 1959.

12. Rose, Joan B., C.P. Gerba, Shri N. Singh, G.A. Toranzos, and Bruce Keswick. "Isolating Viruses from Finished Water." **J.AWWA,** 78(1):56-61, 1986.

13. Cleasby, J.L., D.J. Hilmoe, and C.J. Dimitracopoulos. "Slow-Sand and Direct In-Line Filtration of a Surface Water." **J.AWWA,** p. 44, December, 1984.

14. Karlin, R.J., and R.S. Hopkins. "Engineering Defects Associated with Colorado Giardiasis Outbreaks, June, 1980 to June, 1982." In: Proceedings of the American Water Works Association 1983 Annual Conference.

15. Cleasby, M.L., D.W. Hilmoe, C.J. Dimitracopoulos, and L.M. Diaz-Bossio. "Effective Filtration Methods for Small Water Supplies." EPA-600/2-84-088, U.S. Environmental Protection Agency, May, 1984. NTIS: PB-84-187905.

16. Hendricks, D.W., W.D., Bellamy, and M. Al-Ani. "Removal of Giardia Cysts by Filtration." In: Proceedings of the 1984 Specialty Conference, Environmental Engineering, ASCE, June, 1984.

17. Letterman, R.D., and T.R. Cullen, Jr. "Slow Sand Filter Maintenance Costs and Effects on Water Quality." EPA-600/2-85-056, U.S. Environmental Protection Agency, Cincinnati, OH, May, 1985. NTIS: PB-85-199669/AS.

18. Robeck, G.R., et al. "Effectiveness of Water Treatment Processes in Virus Removal." **J.AWWA,** Vol. 54, p. 1265, 1962.

19. Logsdon, G.S., J.M. Symonds, R.L. Hoye, Jr., and M.M. Arozarena. "Alternative Filtration Methods for Removal of Giardia Cysts and Cyst Model." **J.AWWA,** Vol. 73, pp. 111-118, February, 1981.

SECTION II (Continued)

20. Logsdon, G. "Water Filtration for Asbestos Fiber Removal." EPA-600/2-79-206, U.S. Environmental Protection Agency, MERL, Cincinnati, OH, December, 1979.

21. Tate, C.H., and R.R. Trussell. "Optimization of Turbidity Removal by Use of Particle Counting for Developing Plant Design Criteria." In: Proceedings of AWWA Technology Conference, San Diego, CA, p. 1, December, 1976.

22. Schleppenbach, F.X. "Water Filtration at Duluth." EPA-600/2-84-083, report prepared by the Duluth Water and Gas Department under Grant No. S-804221 from U.S. Environmental Protection Agency, MERL, Cincinnati, OH, April, 1984. NTIS: PB-84-177807.

23. Kavanaugh, M.C., C.H. Tate, A.R. Trussell, R.R. Trussell, and Gordon Treweek. "Use of Particle Size Distribution Measurements for Selection and Control of Solid/Liquid Separation Processes." In: **Particulates in Water,** M.C. Kavanaugh and J.O. Leckie, eds. Advances in Chemistry Series 189, American Chemical Society, Washington, D.C., 1980.

24. Hudson, H.E. "High Quality Water Production and Viral Disease." **J.AWWA,** p. 1265, October, 1962.

25. Leong, L.Y.C. "Removal and Inactivation of Viruses by Treatment Processes for Potable Water and Wastewater - A Review." **Water Science Technology,** 15:91-114.

26. Gerba, Charles P. "Strategies for the Control of Viruses in Drinking Water." University of Arizona, Tucson, AZ, Summer, 1984.

27. Payment, Pierre, Michel Trudel, and Robert Plante. "Elimination of Viruses and Indicator Bacteria at Each Step of Treatment During Preparation of Drinking Water at Seven Water Treatment Plants." **Applied and Environmental Microbiology,** 49(6):1418-1428, June, 1985.

SECTION II (Continued)

28. Stetler, R.E., R.L. Ward, and S.C. Waltrip. "Enteric Virus and Indicator Bacteria Levels in a Water Treatment System Modified to Reduce Trihalomethane Production." **Applied and Environmental Microbiology**, 47:319-324, 1983.

29. Gerba, C.P., J.B. Rose, G.A. Toranzos, S.N. Singh, L.M. Kelley, B. Keswick, and H.L. DuPont. "Virus Removal During Conventional Drinking Water Treatment." EPA-600/2-85-017, U.S. Environmental Protection Agency, Cincinnati, OH, September, 1985.

30. AWWA Committee. "Viruses in Drinking Water." **J.AWWA,** 71(8):441-444, 1979.

31. Bull, R.J. "Toxicological Problems Associated with Alternative Methods of Disinfection." **J.AWWA,** pp. 642-648, December, 1982.

32. Eaton, J.W., C.F. Koplin, H.S. Swofford, C.M. Kjellstrand, and H.S. Jacob. "Chlorinated Urban Water: A Cause of Dialysis Induced Hemolytic Anemia." **Science,** 181:463-464, 1973.

33. U.S. Environmental Protection Agency. "Trihalomethanes in Drinking Water: Sampling, Analysis, Monitoring and Compliance." EPA-570/9-83-002, U.S. Environmental Protection Agency, Office of Drinking Water, Washington, D.C., August, 1983.

34. Association for the Advancement of Medical Instrumentation. "Proposed Standard for Hemodialysis Systems." AAMI RD5 - 1981, June, 1981.

35. Craun, G.F. "Outbreaks of Waterborne Disease in the United States: 1971-1978." **J.AWWA,** 73(7):360-369, 1981.

36. Kirner, J.C., J.D. Littler, and L.A. Angelo. "A Waterborne Outbreak of Giardiasis in Camas, Washington." **J.AWWA,** 70(1):35-39, 1978.

37. Lippy, E.C. "Tracing a Giardiasis Outbreak at Berlin, New Hampshire." **J.AWWA,** 70(9):512-520, 1978.

SECTION II (Continued)

38. Lin, S.D. "Giardia Lamblia and Water Supply." **J.AWWA,** p. 40, February, 1985.

39. American Water Works Association. "Quality Goals for Public Water, Statement of Policy." **J.AWWA,** p. 1317, 1968.

40. Culp, R.L., et al. "Increasing Water Treatment Capacity with Minimum Additions." **Public Works,** August, 1976.

41. Harris, W.L. "High-Rate Filter Efficiency." **J.AWWA,** p. 515, 1970.

42. Regli, S., A. Amirtharajah, J. Hoff, and P. Berger. "Treatment for Control of Waterborne Pathogens: How Safe is Safe Enough?" In: Proceedings, 3rd Conference on Progress in Chemical Disinfection, G.E. Janauer (Ed.), April 3-5, 1986. State University of New York, Binghampton, NY, in press.

43. Greenberg, A.E., "Public Health Aspects of Alternative Water Disinfectants." **J.AWWA**, 73(1):31-33, 1981.

44. Amirtharajah, A. "Variance Analyses and Criteria for Treatment Regulations." **J.AWWA,** 78:34-49, March, 1986.

45. McCabe, L.J., M.M. Symonds, R.D. Lee, and G.G. Robeck. "Survey of Community Water Supply Systems." **J.AWWA,** 62(11):670-687, 1970.

46. Geldreich, E.E., H.D. Nash, D.J. Reasoner, and R.H. Taylor. "The Necessity of Controlling Bacterial Populations in Potable Waters: Community Water Supply." **J.AWWA,** 64(9):596-602, 1972.

47. American Water Works Association. "Committee Report: Microbiological Considerations for Drinking Water Regulation Revisions." **J.AWWA,** 79(5), 1987.

SECTION III

1. Amirtharajah, A. "Variance Analyses and Criteria for Treatment Regulations." **J.AWWA,** 78:34-49, March, 1986.

SECTION III (Continued)

2. Stetler, R.E., R.L. Ward, and S.C. Waltrip. "Enteric Virus and Indicator Bacteria Levels in a Water Treatment System Modified to Reduce Trihalomethane Production." **Applied and Environmental Microbiology**, 47:319-324, 1983.

3. Taylor, E.W. "Forty-fifth Report on the Results of the Bacteriological, Chemical, and Biological Examination of the London Waters for the Years 1971-73." Metropolitan Water Board, London, England, no date.

4. Payment, Pierre, Michel Trudel, and Robert Plante. "Elimination of Viruses and Indicator Bacteria at Each Step of Treatment During Preparation of Drinking Water at Seven Water Treatment Plants." **Applied and Environmental Microbiology**, 49(6):1418-1428, June, 1985.

5. DeWalle, F.B., J. Engeset, and W. Lawrence. "Removal of Giardia Lamblia Cysts by Drinking Water Plants." EPA-600/S2-84-069, U.S. Environmental Protection Agency, MERL, Cincinnati, OH, May, 1984.

6. Al-Ani, M., J.M. McElroy, C.P. Hibler, and D.W. Hendricks. "Filtration of Giardia Cysts and Other Substances, Vol 3: Rapid Rate Filtration." EPA-600/2-85-027, U.S. Environmental Protection Agency, MERL, Cincinnati, OH, April, 1985. NTIS: PB-85-194645/AS.

7. Lange, K.P., W.D. Bellamy, and D.W. Hendricks. "Filtration of Giardia Cysts and Other Substances, Vol. 1: Diatomaceous Earth Filtration." EPA-600/2-84-114, U.S. Environmental Protection Agency, MERL, Cincinnati, OH, June, 1984. NTIS: PB-84-212703.

8. Bellamy, W.D., G.P. Silverman, and D.W. Hendricks. "Filtration of Giardia Cysts and Other Substances, Vol. 2: Slow Sand Filtration." EPA-600/2-85-026, U.S. Environmental Protection Agency, MERL, Cincinnati, OH, April, 1985. NTIS: PB-85-191633/AS.

9. Logsdon, G.S. "Comparison of Some Filtration Processes Appropriate for Giardia Cyst Removal." Presented at the Calgary Giardia Conference, Calgary, Alberta, Canada, February 23-25, 1987.

SECTION III (Continued)

10. Cleasby, J.L., D.J. Hilmoe, and C.J. Dimitracopoulos. "Slow-Sand and Direct In-Line Filtration of a Surface Water." **J.AWWA,** p. 44, December, 1984.

11. Cleasby, M.L., D.W. Hilmoe, C.J. Dimitracopoulos, and L.M. Diaz-Bossio. "Effective Filtration Methods for Small Water Supplies." EPA-600/2-84-088, U.S. Environmental Protection Agency, May, 1984. NTIS: PB-84-187905.

12. Logsdon, G.S., et al. "Evaluating Sedimentation and Various Filter Media for Removal of Giardia Cysts." **J.AWWA,** p. 61, February, 1985.

13. Lin, S.D. "Giardia Lamblia and Water Supply." **J.AWWA,** p. 40, February, 1985.

14. Personal communication. Ron Hartman, Chief Operator, Moffat Water Treatment Plant, Denver Water Board, Denver, Colorado.

15. Sequeira, James G., L. Harry, S.P. Hansen, and R.L. Culp. "Pilot Plant Filtration Tests at the American River Water Treatment Plant." **Public Works Works,** January, 1983.

16. Culp, R.L. "Direct Filtration." **J.AWWA,** p. 375-378, July, 1977.

17. Committee report. "The Status of Direct Filtration." **J.AWWA,** p. 405, July, 1980.

18. Hendricks, D.W., W.D., Bellamy, and M. Al-Ani. "Removal of Giardia Cysts by Filtration." In: Proceedings of the 1984 Specialty Conference, Environmental Engineering, ASCE, June, 1984.

19. McCormick, Richard Ford, and Paul H. King. "Factors that Affect Use of Direct Filtration in Treating Surface Waters." **J.AWWA,** p. 234, May, 1982.

20. Westerhoff, Garret P., Alan F. Hess, and M.J. Barnes. "Plant-Scale Comparison of Direct Filtration Versus Conventional Treatment of a Lake Erie Water." **J.AWWA,** p. 148, March, 1980.

SECTION III (Continued)

21. Logsdon, G.S., R.M. Clark, and C.H. Tate. "Direct Filtration Treatment Plants." **J.AWWA,** 72(3):134-147, March, 1980.

22. Logsdon, G.S., G.L. Evavold, J.L. Patton, and J. Watkins, Jr. "Filter Plant Design for Asbestos Fiber Removal." **ASCE J. Environmental Engineering,** 109(4):900-914, August, 1983.

23. Personal communication. Jerry Blinn, Chief Operator, Plumas County Flood Control and Water Conservation District, Quincy, California, June, 1985.

24. James M. Montgomery, Consulting Engineers, Inc. **Water Treatment Principles and Design.** John Wiley and Sons, Inc., New York, NY, 1985.

25. Malina, J.F., Jr., B.D. Moore, and J.L. Marshall. "Poliovirus Removal by Diatomaceous Earth Filtration." Center for Research in Water Resources, The University of Texas, Austin, TX, 1972.

26 Baumann, E.R. **Diatomite Filtration of Potable Water, Water Quality and Treatment, A Handbook of Public Water Supplies.** American Water Works Association, McGraw-Hill Book Co., NY, 1971.

27. Pyper, Gordon R. "Slow Sand Filter and Package Treatment Plant Evaluation Operating Costs and Removal of Bacteria, Giardia, and Trihalomethanes." EPA-600/2-85-052, U.S. Environmental Protection Agency, Cincinnati, OH, April, 1985.

28. Culp/Wesner/Culp. "Technical Guidelines for Public Water Systems." EPA Contract No. 68-01-2971, U.S. Environmental Protection Agency, Cincinnati, OH, June, 1975.

29. Letterman, R.D., and T.R. Cullen, Jr. "Slow Sand Filter Maintenance Costs and Effects on Water Quality." EPA-600/2-85-056, U.S. Environmental Protection Agency, Cincinnati, OH, May, 1985. NTIS: PB-85-199669/AS.

30. Personal communication. Charles R. Beer, Superintendent of Water Treatment, Denver Water Board, Denver, Colorado, May, 1985.

SECTION III (Continued)

31. Slezak, Lloyd A., and R.C. Sims. "The Application and Effectiveness of Slow Sand Filtration in the United States." **J.AWWA,** Vol. 76, No. 12, p. 38, December, 1984.

32. Morand, James M., and Mather J. Young. "Performance Characteristics of Package Water Treatment Plants, Project Summary." EPA-600/S2-82-101, U.S. Environmental Protection Agency, MERL, Cincinnati, OH, March, 1983.

33. Morand, James M., Craig R. Cobb, Robert M. Clark, and Richard, G. Stevie. "Package Water Treatment Plants, Vol. 1, A Performance Evaluation." EPA-600/2-80-008a, U.S. Environmental Protection Agency, MERL, Cincinnati, OH, July, 1980.

34. Association of State Drinking Water Administrators. "Survey to Support Analysis of Proposed Regulations Concerning Filtration and Disinfection of Public Drinking Water Supplies." EPA Grant T 901534-01-0. Arlington, VA, September 5, 1986.

35. Horn, J.B., and D.W. Hendricks. "Removals of Giardia Cysts and Other Particles From Low Turbidity Waters Using the Culligan Multi-Tech Filtration System." Engineering Research Center, Colorado State University, unpublished, 1986.

36. Robeck, G.G., N.A. Clarke, and K.A. Dostal. "Effectiveness of Water Treatment Processes in Virus Removal." **J.AWWA,** 54(10):1275-1290, 1962.

SECTION IV

1. Fair, G.M., et al. **Water Supply and Waste Disposal.** John Wiley & Sons, New York, NY, p. 552, 1959.

2. Hoff, J.C. "Inactivation of Microbial Agents by Chemical Disinfectants." EPA/600/S2-86/067, U.S. Environmental Protection Agency, Water Engineering Research Laboratory, Cincinnati, OH, September, 1986.

SECTION IV (Continued)

3. Berger, P.S., and Y. Argaman. "Assessment of Microbiology and Turbidity Standards for Drinking Water." U.S. Environmental Protection Agency Report, EPA-570/9-83-001, Washington, D.C., 1983.

4. Rice, E.W., J.C. Hoff, and F.W. Schaeffer. "Inactivation of *Giardia* Cysts by Chlorine." **Applied and Environmental Microbiology**, 43:250-251, 1982.

5. Wickramanayake, G.B., A.J. Rubin, and O.J. Sproul. "Effects of Ozone and Storage Temperature on *Giardia* Cysts." **J.AWWA**, 77(8):74-77, 1985.

6. Wickramanayake, G.B., A.J. Rubin, and O.J. Sproul. "Inactivation of *Giardia Lamblia* Cysts with Ozone." **Applied and Environmental Microbiology**, 48(3):671, 1984.

7. Meyer, E.A. "Disinfection of *Giardia Muris* in Chloraminated Water." Unpublished. Research conducted for City of Portland, Oregon, 1982.

8. Kruse, C.W., V.P. Olivieri, and K. Kawata. "The Enhancement of Viral Inactivation by Halogens." **Water and Sewage Works**, p. 187, June, 1971.

9. Culp, G.L., and R.L. Culp. **New Concepts in Water Purification.** Van Nostrand Reinhold Company, New York, NY, 1974.

10. Lippy, E.C., and G. Logsdon. "Where Does Giardiasis Occur and Why?" In: Proceedings of the 1984 Specialty Conference of the Environmental Engineering Division, ASCE, p. 226, University of Southern California, Los Angeles, CA, June 25-27, 1984.

11. Baumann, E.R., and D.D. Ludwig. "Free Available Chlorine Residuals for Small Nonpublic Water Supplier." **J.AWWA**, p. 1379, November, 1962.

12. Liu, O.C., H.R. Seraichekas, E.W. Akin, D.A. Brashear, E.L. Katz, and W.L. Hill, Jr. "Relative Resistance of Twenty Human Enteric Viruses to Free Chlorine in Potomac Water." In: Virus and Water Quality: Occurrence and Control. 13th Water Quality Conference, University of Illinois, Urbana-Champaign, 1971.

SECTION IV (Continued)

13. Clarke, N.A., G. Berg, P.W. Kabler, and L.L. Chang. "Human Enteric Viruses in Water: Source, Survival, and Removability." International Conference on Water Pollution Research, Landar, September, 1962.

14. Hoff, J.C., E.W. Rice, and F.W. Schaefer III. "Disinfection and the Control of Waterborne Giardiasis." In: Proceedings of the 1984 Specialty Conference, Environmental Engineering, ASCE, June, 1984.

15. Jarroll, E.L., A.K. Bingham, and E.A. Meyer. "Effect of Chlorine in Giardia Lamblia Cyst Viability." **Applied and Environmental Microbiology**, 41:483-487, February, 1981.

16. Scarpino, P.V., G. Berg, S.L. Chang, D. Dahling, and M. Lucas. "A Comparative Study of the Inactivation of Viruses in Water by Chlorine." **Water Research**, 6, pp. 959-965, 1972.

17. Snow, W.B. "Recommended Chlorine Residuals for Military Water Supplies." **J.AWWA**, 48:1510-1514, 1956.

18. DeWalle, F.B., and C.R.E. Jansson. "Inactivation of Giardia by Chlorine and UV." In: Proceedings of the Seminar on Giardiasis and Public Water Supplies, British Columbia Water and Waste Association, Richmond, B.C., November 22, 1983.

19. Lippy, E.C. "Chlorination to Prevent and Control Waterborne Diseases." **J.AWWA**, 78(1):49-52, 1986.

20. White, G.C. **Handbook of Chlorination.** Van Nostrand Reinhold Company, New York, NY, 1972.

21. Olson, K.E. "An Evaluation of Low Chlorine Concentrations on Giardia Cyst Viability." USDA Forest Service, Equipment Development Center, San Dimas, CA, January, 1982.

22. Blaser, M.J., R.G. West, W.L. Wang, and J.C. Hoff. "Control of Waterborne Campylobacter Jejuni by Chlorine Disinfection." In: Proceedings of the AWWA Water Quality Technology Conference, Norfolk, VA, December 4-7, 1983.

SECTION IV (Continued)

23. Scarpino, P.V. Unpublished data, 1978.

24. Yee, R.B., and R.M. Wadowsky. "Multiplication of Legionella Pneumophila in Unsterilized Tap Water." **Applied and Environmental Microbiology**, 43:1330-1334, 1982.

25. Skaliy, P., T.A. Thompson, G.W. Gorman, G.K. Morris, V. McEachern, and D.C. Mackel. "Laboratory Studies of Disinfectants Against Legionella Pneumophila." **Applied and Environmental Microbiology**, 40:697-700, 1980.

26. Fliermans, C.B., G.E. Bettinger, and A.W. Fynsk. "Treatment of Cooling Systems Containing High Levels of Legionella Pneumophila." **Water Resources**, 16:903-909, 1982.

27. England, A.C., III, D.W. Fraser, G.F. Mallison, D.C. Mackel, P. Skaliy, and C.W. Gorman. "Failure of Legionella Pneumophila Sensitivities to Predict Culture Results from Disinfection-Treated Air-Conditioned Cooling Towers." **Applied and Environmental Microbiology**, 43:240-244, 1982.

28. Kuchta, J.M., S.J. States, J.E. McGlaughlin, J.H. Overmeyer, R.M. Wadowsky, A.M. McNamara, R.S. Wolford, and R.B. Yee. "Enhanced Chlorine Resistance of Tap Water-Adapted Legionella Pneumophila as Compared with Agar Medium-Passaged Strains." **Applied and Environmental Microbiology**, 50(1):21-26, July, 1985.

29. Fisher-Hoch, S.P., et al. "Investigations and Control of an Outbreak of Legionnaires' Disease in a District General Hospital." **The Lancet**, 2:932, 1981.

30. Hoff, J.C., and E.E. Geldreich. "Adequacy of Disinfection for Control of Newly Recognized Waterborne Pathogens." In: Proceedings of the AWWA Water Quality Technology Conference, Nashville, Tennessee, December, 5-8, 1982.

31. Keswick, B.H., T.K. Satterthwaite, P.C. Johnson, H.L. DuPont, S.L. Secor, J.A. Bitsura, W. Gary, and J.C. Hoff. "Inactivation of Norwalk Virus in Drinking Water by Chlorine." **Applied and Environmental Microbiology**, 50(2):261-264, August, 1985.

SECTION IV (Continued)

32. Logsdon, G.S., and E.C. Lippy. "The Role of Filtration in Preventing Waterborne Disease." **J.AWWA**, p. 649, December, 1982.

33. Amirtharajah, A. "Variance Analyses and Criteria for Treatment Regulations." **J.AWWA**, 78:34-49, March, 1986.

34. Keswick, B.H., C.P. Gerba, H.L. Dupont, and J.B. Rose. "Detection of Enteric Viruses in Treated Drinking Water." **Applied and Environmental Microbiology**, 47(6):1290-1294, 1984.

35. Payment, Pierre, Michel Trudel, and Robert Plante. "Elimination of Viruses and Indicator Bacteria at Each Step of Treatment During Preparation of Drinking Water at Seven Water Treatment Plants." **Applied and Environmental Microbiology**, 49(6):1418-1428, June, 1985.

36. Stetler, R.E., R.L. Ward, and S.C. Waltrip. "Enteric Virus and Indicator Bacteria Levels in a Water Treatment System Modified to Reduce Trihalomethane Production." **Applied and Environmental Microbiology**, 47:319-324, 1983.

37. O'Connor, J.T., L. Hemphill, and C.D. Reach, Jr. "Removal of Virus From Public Water Supplies." EPA-600/2-82-024, U.S. Environmental Protection Agency, Cincinnati, OH, 160 pp., August, 1982.

38. Ridenour, G.M., and R.S. Ingols. "Bactericidal Properties of Chlorine Dioxide." **J.AWWA**, Vol. 39, 1947.

39. Bernarde, M.A., B.M. Israel, V.P. Olivieri, and M.L. Granstrom. "Efficiency of Chlorine Dioxide as a Bactericide." **Applied and Environmental Microbiology**, Vol. 13, 1965.

40. Chen, Y.S.R., O.J. Sproul, and A.J. Rubin. "Inactivation of Naegleria Gruberi Cysts by Chlorine Dioxide." U.S. Environmental Protection Agency, Grant R808150-02-0, Department of Civil Engineering, Ohio State University, 1984.

SECTION IV (Continued)

41. Symons, J.M., A.A. Stevens, R.M. Clark, E.E. Geldreich, O.T. Love, Jr., and J. DeMarco. "Treatment Techniques for Controlling Trihalomethanes in Drinking Water." EPA-600/2-81-156, U.S. Environmental Protection Agency, Cincinnati, OH, 1981.

42. Cronier, S., et al. **Water Chlorination, Volume 2.** Jolley, R.L., et al, editors, Ann Arbor Science Publishers, Inc., Ann Arbor, MI, 1978.

43. Rubin, A.J., J.P. Engel, and O.J. Sproul. "Disinfection of Amoebic Cysts in Water with Free Chlorine." **J.WPCF,** 55:1174-1182, 1983.

44. Wickramanayake, G.B., A.J. Rubin, and O.J. Sproul. "Inactivation of *Naegleria* and *Giardia* Cysts in Water by Ozonation." **J.WPCF**, 56(8):983-988, 1984.

45. Eaton, J.W., C.F. Koplin, H.S. Swofford, C.M. Kjellstrand, and H.S. Jacob. "Chlorinated Urban Water: A Cause of Dialysis Induced Hemolytic Anemia." **Science,** 181:463-464, 1973.

46. Bull, R.J. "Health Effects of Alternate Disinfectants and Their Reaction Products." **J.AWWA,** 72:299, 1980.

47. Wolfe, R.L., N.R. Ward, and B.H. Olson. "Inorganic Chloramines as Drinking Water Disinfectants: A Review." **J.AWWA,** 76(5):74-88, 1984.

48. Hoff, J.C., and E.E. Geldreich. "Comparison of the Biocidal Efficiency of Alternative Disinfectants." **J.AWWA**, 73(1):40-44, 1981.

49. Culp/Wesner/Culp. "Evaluation of Treatment Effectiveness for Reducing Trihalomethanes in Drinking Water." EPA Contract No. 68-01-6292, U.S. Environmental Protection Agency, Cincinnati, OH, July, 1983.

50. Shull, K.E. "Experience With Chloramines as Primary Disinfectants." **J.AWWA,** 73(2):101-104, 1981.

SECTION IV (Continued)

51. Olivieri, V.P., M.C. Snead, C.W. Kruse, and K. Kawata. "Stability and Effectiveness of Chlorine Disinfectants in Water Distribution Systems." EPA-600/S2-84-011, U.S. Environmental Protection Agency, March, 1984.

52. Rice, R.G., C.M. Robson, G.W. Miller, and A.G. Hill. "Uses of Ozone in Drinking Water Treatment." **J.AWWA,** 73(1):44-57, 1981.

53. Personal communication. Rip G. Rice, Inc., September, 1984.

54. Kessel, J.F., et al. "The Cysticidal Effect of Chlorine and Ozone on Cysts of Entamoeba Histolytica." **Am. J. Trop. Med.,** 24:117, 1944.

55. Joost, R.D., B.W. Long, and L. Jackson. "Using Ozone as A Primary Disinfectant for the Tucson C.A.P. Water Treatment Plant." Presented at the 1988 Annual Meeting of the International Ozone Conference, Monroe, Michigan, April, 1988.

56. LePage, W.L. "The Anatomy of an Ozone Plant." **J.AWWA**, p. 105, February, 1981.

57. Martin, Brian C., and John C. Goble. "Ozone Assists in Treating a Lake Water - A Case History." Presented at the California-Nevada AWWA Section Meeting, October, 1984.

58. National Research Council (National Academy of Sciences). **Drinking Water and Health, Volume 3.** National Academy Press, pp. 302-309, 1980.

59. Black, A.P., W.C. Thomas, Jr., R.N. Kinman, W.P. Bonner, M.A. Keirn, J.J. Smith, Jr., and A.A. Jabero. "Iodine for the Disinfection of Water." **J.AWWA,** 60(1):69, 1968.

60. Karalekas, P.C., Jr., L.N. Kuzminski, and T.H. Feng. "Recent Developments in the Use of Iodine for Water Disinfection." **Journal New England Water Works Association,** p. 152, June, 1970.

SECTION IV (Continued)

61. McKee, J.E., C.J. Brokaw, and R.T. McLaughlin. "Chemical and Colicidal Effects of Halogens in Sewage." **J.WPCF,** p. 795, August, 1960.

62. Cook, B. "Iodine Dispenser for Water Supply Disinfection." USDA Forest Service, San Dimas Equipment Development Center, CA, January, 1976.

63. National Research Council (National Academy of Sciences). **Drinking Water and Health, Vol. 2.** National Academy Press, pp. 114-115, 1980.

64. Ibid, p. 189, 1980.

65. Rice, E.W., and J.C. Hoff. "Inactivation of Giardia Lamblia Cysts by Ultraviolet Radiation." **Applied and Environmental Microbiology,** p. 546, September, 1981.

66. Kawabata, T., and T. Harada. "The Disinfection of Water by the Germicidal Lamp." **J. Illumination Society,** 36:89, 1959.

67. Huff, C.B., H.F. Smith, W.D. Boring, and N.A. Clarke. "Study of Ultraviolet Disinfection of Water and Factors in Treatment Efficiency." **Public Health Reports,** 80:695-705, August, 1965.

68. Kelner, A. "Effect of Visible Light on the Recovery of Streptomyces griseus conidia From Ultraviolet Irradiation Injury." In: Proceedings of the National Academy of Sciences, U.S., 35:73, 1949.

69. Lawrence, C.A., and S.S. Block. **Disinfection, Sterilization and Preservation.** Lea and Febiger, Philadelphia, PA, 1968.

70. Scheible, O.K., A. Forndran, and W.M. Leo. "Pilot Investigation of Ultraviolet Wastewater Disinfection at the New York City Port Richmond Plant." Paper presented at Second National Symposium at Municipal Wastewater Disinfection. In Symposium Proceedings, EPA-600/9-83-009, Cincinnati, OH, July, 1983.

SECTION IV (Continued)

71. Witherell, L.E., R.L. Solomon, and K.M. Stone. "Ozone and Ultraviolet Radiation Disinfection for Small Community Water Systems." EPA-600/2-79-060, U.S. Environmental Protection Agency, Municipal Environmental Research Laboratory, Cincinnati, OH, July, 1979.

72. Symons, J.M., J.K. Carswell, R.M. Clark, P. Dorsey, E.E. Geldreich, W.P. Heffernan, J.C. Hoff, O.T. Love, L.J. McCabe, and A.A. Stevens. "Ozone, Chlorine Dioxide, and Chloramines as Alternatives to Chlorine for Disinfection of Drinking Water: State of the Art." U.S. Environmental Protection Agency, Water Supply Research Division, Cincinnati, OH, 1977.

73. Committee report. "Viruses in Water." **J.AWWA**, p. 491, October, 1969.

74. Sproul, O.J. "Protected Watersheds are Often a Health Hazard." College of Engineering and Physical Sciences, University of New Hampshire, Durham, NH Presented at the AWWA Annual Meeting in Washington, D.C., 1985.

75. Katzenelson, E., B. Kletter, and H.I. Shuval. "Inactivation Kinetics of Viruses and Bacteria in Water by Use of Ozone." **J.AWWA**, 66(12):725, 1974.

76. Roy, D., R.S. Engelbrecht, and E.S.K. Chian. "Comparative Inactivation of Six Enteroviruses by Ozone." **J.AWWA**, 74(12):660, 1982.

77. AWWA Committee. "Viruses in Drinking Water." **J.AWWA**, 71(8):441-444, 1979.

78. Payment, P., et al. "Relative Resistance to Chlorine of Poliovirus and Coxsackievirus Isolates from Environmental Source and Drinking Water." **Applied and Environmental Microbiology**, 49(4):981-983, April 1985.

79. Rubin, A.J. U.S. Environmental Protection Agency, Contract No. CR-812238, Quarterly Report, December, 1985, unpublished.

SECTION IV (Continued)

80. Muraca, P., J.E. Stout, and V.L. Yee. "Comparative Assessment of Chlorine, Heat, Ozone, and UV Light for Killing Legionella pneumophila Within a Model Plumbing System." **Applied and Environmental Microbiology,** 53(2):447-453, 1987.
81. Katzenelson, E., B. Kletter, and H.I. Shuval. "Inactivation Kinetics of Viruses and Bacteria in Water by Use of Ozone." **J.AWWA,** 66(12):725, 1974.

82. Roy, D., R.S. Engelbrecht, and E.S.K. Chian. "Comparative Inactivation of Six Enteroviruses by Ozone." **J.AWWA,** 74(12):660, 1982.

SECTION V

1. Malcolm Pirnie, Inc. "Analysis of Plant Capacities and Average Flow Rates." Letter report prepared for the U.S. Environmental Protection Agency, Office of Drinking Water, Washington, D.C., January 6, 1986.

2. Malcolm Pirnie, Inc., and Consumers/Pirnie Utility Services Company. "Evaluation of Costs Associated With Turbidity Removal Systems in Small Water Companies." Draft report prepared for the U.S. Environmental Protection Agency, July, 1986.

3. Lippy, E.C., and S.C. Waltrip. "Waterborne Disease Outbreaks--1946-1980: A Thirty-Five Year Perspective." **J.AWWA,** p. 60, February, 1984.

4. Culp/Wesner/Culp, Consulting Engineers. "Survey and Evaluation of 80 Public Water Systems in Wyoming." Prepared in three volumes for the U.S. Environmental Protection Agency (Region VIII) and the National Environmental Health Association, January, 1986.

5. U.S. Environmental Protection Agency. "Small System Water Treatment Symposium." EPA-570/9-79-021, Report of Symposium Proceedings, Cincinnati, OH, November 28-29, 1978.

SECTION V (Continued)

6. Morand, James M., Craig R. Cobb, Robert M. Clark, and Richard, G. Stevie. "Package Water Treatment Plants, Vol. 1, A Performance Evaluation." EPA-600/2-80-008a, U.S. Environmental Protection Agency, MERL, Cincinnati, OH, July, 1980.

7. Association of State Drinking Water Administrators. "Survey to Support Analysis of Proposed Regulations Concerning Filtration and Disinfection of Public Drinking Water Supplies." EPA Grant T 901534-01-0. Arlington, VA, September 5, 1986.

8. Hollingsworth, J.E., R.E. Holt, and W. Hults. "Filter Press Traps Giardia Cysts." Manufacturer's Forum, **Water World News,** September-October, 1985.

9. Kutadyne Products Incorporated, product brochure.

10. Long, W.R. "Evaluation of Cartridge Filters for the Removal of Giardia Lamblia Cyst Models from Drinking Water Systems." **Journal of Environmental Health,** 45(5):220-225, March/April, 1983.

11. Hibler, C.P. "Evaluation of the 3M Filter 124A in the FS-SR 122 Type 316 s/s #150 Housing for Removal of Giardia Cysts." Department of Pathology, Colorado State University, unpublished, 1986.

12. Cook, B. "Iodine Dispenser for Water Supply Disinfection." USDA Forest Service, San Dimas Equipment Development Center, CA, January, 1976.

13. Olson, K.E. "Evaluation of Erosion Feed Chlorinators." EPA/600/S2-85-126, U.S. Environmental Protection Agency, Water Engineering Research Laboratory, Cincinnati, OH, December, 1985.

SECTION VI

1. Culp/Wesner/Culp. "Estimation of Small System Water Treatment Costs." EPA Contract No. 68-03-3093, U.S. Environmental Protection Agency, Cincinnati, OH. In publication.

SECTION VI (Continued)

2. Culp/Wesner/Culp. "Estimating Water Treatment Costs, Volume 2: Cost Curves Applicable to 1 to 200 mgd Treatment Plants." EPA-600/2-79-162b, U.S. Environmental Protection Agency, Cincinnati, OH, August, 1979.

3. Malcolm Pirnie, Inc. "Analysis of Plant Capacities and Average Flow Rates." Letter report prepared for the U.S. Environmental Protection Agency, Office of Drinking Water, Washington, D.C., January 6, 1986.

APPENDIX D

1. American Water Works Association, Government Affairs Office. "Surface Water Treatment Rule Evaluation Project." Final Report, December, 1987.

Appendix A: Groundwater Disinfection Costs

Well water supplies which have no facilities for disinfection are helpless in the face of microbial contamination, and they have no means of providing residual protection against contamination of water in transit or in storage in the distribution system. Unless site-specific conditions significantly mitigate such concerns, disinfection of groundwaters appears warranted. Several states already have mandatory disinfection requirements.

Viruses have been identified as the key organism of concern for groundwater sources because they are capable of traveling great distances in groundwater. Among many similar examples, Gerba notes that in fractured rock and similar soils, viruses have been observed to travel as far as 1,600 meters, while in gravel or sandy soils they have been found as far as 900 meters from their point of origin (Ref. 26, Section II).

Since viruses can be effectively inactivated by disinfection, the following tables present representative costs of disinfection by several methods. These costs have been developed on the assumption that the groundwater source is clear (NTU $<$ 1.0) and unpolluted by dissolved organic or inorganic chemical compounds. Actual costs for a specific sites would of course have to be modified by conditions at that site.

Costs in these tables differ from those for surface water disinfection facilities for several reasons. First, system capacities are different from those for surface water systems, based on statistical analysis of system size data collected during this project. Second, the following costs for groundwater systems include detention facilities for disinfectant contact, where disinfection costs in Section VI (surface water) do not, since it has been assumed that no detention facilities exist when disinfection is added to a well supply source, whereas disinfection added to a filtration plant could, as one alternative, use the clearwell storage unit for detention. Construction costs for detention structures were minmized, however, by using the assumptions described in the footnotes of each table.

TABLE A-1. SUMMARY OF GROUNDWATER DISINFECTION COSTS

DISINFECTION METHOD	COST OF TREATMENT, ¢/1,000 GAL											
	Size, Category											
	1	2	3	4	5	6	7	8	9	10	11	12
	0.06	0.14	0.31	0.96	3.06	7.52	15.4	25.3	44.2	124	465	1,505
	0.013	0.045	0.133	0.40	1.30	3.25	6.75	11.5	20.0	55.5	205	650
Chlorine[1]	66.9[2]	22.9	12.0	4.7	3.1	1.6	1.1	0.9	0.8	0.6	0.4	0.3
	30.2[2]	19.4	8.7	6.2	5.4	3.3	2.5	2.0	1.7	1.3	1.0	0.8
	97.1[2]	42.3	20.7	10.9	8.5	4.9	3.6	2.9	2.5	1.9	1.4	1.1
Ozone[3]	165.1	61.6	31.6	15.5	7.5	5.0	4.0	3.5	3.1	2.1	1.4	1.2
	-	-	-	-	-	-	-	-	-	-	-	-
	165.1	61.6	31.6	15.5	7.5	5.0	4.0	3.5	3.1	2.1	1.4	1.2
Chlorine Dioxide[4]	321.3	96.9	45.3	16.8	6.0	3.1	2.1	1.6	1.4	0.9	0.6	0.4
	19.7	10.0	4.7	3.3	3.7	2.4	1.6	1.3	1.1	0.8	0.5	0.4
	341.0	106.9	50.0	20.1	9.7	5.5	3.7	2.9	2.5	1.7	1.1	0.8
Chloramination[5]	116.8	41.8	23.6	9.5	3.9	1.9	1.2	1.0	0.8	0.6	0.3	0.2
	30.2	19.4	8.7	6.2	5.4	3.3	2.5	2.0	1.7	1.3	1.0	0.8
	147.0	61.2	32.3	15.7	9.3	5.2	3.7	3.0	2.5	1.9	1.3	1.0
Ultraviolet Light[6]	61.9	20.2	10.1	7.6								
	-	-	-	-								
	61.9	20.2	10.1	7.6								

[1] Costs include a chlorine dosage of 2.0 mg/l with 30-minute detention time. Chlorine is fed as a hypochlorite solution for Categories 1 through 4; cylinder storage and feed is used for Categories 5 through 10, with on-site tank storage used in Categories 11 and 12. Detention storage is provided by a pressure vessel in Categories 1 and 2, a looped underground pipeline in Categories 3 and 4, and a chlorine contact basin in Categories 5 through 12.

[2] Costs, including both capital and O&M, are shown as follows:

- xx.x - disinfection generation/feed equipment
- yy.y - detention storage facilities
- zz.z - total cost

[3] Includes direct in-line ozone application at a dosage of 1.0 mg/l followed by a 5-minute contact time, assumed to be achieved in the transmission line between the well and the distribution system.

[4] Includes chlorine dioxide at a dose of 2.0 mg/l with a 15-minute detention time. Detention is in pressure vessels in Categories 1 and 2, in looped underground pipelines in Categories 3 and 4, and in chlorine contact basins in Categories 5 through 12.

[5] Costs are for a chlorine dose of 1.5 mg/l, an ammonia dose of 0.5 mg/l, and 30-minute detention. Chlorine is provided as a hypochlorite solution in Categories 1 through 4, by cylinder storage and feed in Categories 5 through 10, and is stored in on-site tanks in Categories 11 and 12. Ammonia is fed as anhydrous ammonia in Categories 1 through 4, and as aqua ammonia in Categories 5 through 12. Detention is in pressure vessels in Categories 1 and 2, in looped underground pipelines in Categories 3 and 4, and in chlorine contact chambers in Categories 5 through 12.

[6] No additional detention storage used beyond that built into the ultraviolet light unit.

TABLE A-2. ESTIMATED COSTS FOR GROUNDWATER DISINFECTION AND DETENTION USING CHLORINE

Category	Plant Capacity, mgd	Average Flow, mgd	Capital Cost, $1,000		O&M Cost, $1,000		Unit Cost, ¢/1,000 gal		Total
			Disinfection	Storage	Disinfection	Storage	Disinfection	Storage	
1	0.06	0.013	10	9	2.0	0.4	66.9	30.2	97.1
2	0.14	0.045	15	22	2.0	0.6	22.9	19.4	42.3
3	0.31	0.133	19	30	3.6	0.7	12.0	8.7	20.7
4	0.96	0.40	27	68	3.7	1.1	4.7	6.2	10.9
5	3.06	1.30	36	216	10.5	0	3.1	5.4	8.5
6	7.52	3.25	50	330	13.2	0	1.6	3.3	4.9
7	15.4	6.75	74	521	18.2	0	1.1	2.5	3.6
8	25.3	11.5	104	732	24.8	0	0.9	2.0	2.9
9	44.2	20.0	163	1,083	36.6	0	0.8	1.7	2.5
10	124	55.5	348	2,264	84.0	0	0.6	1.3	1.9
11	465	205	582	6,229	211	0	0.4	1.0	1.4
12	1,505	650	993	15,483	625	0	0.3	0.8	1.1

Costs include a chlorine dosage of 2.0 mg/l with 30-minute detention time. Chlorine is fed as a hypochlorite solution for Categories 1 through 4; cylinder storage and feed is used for Categories 5 through 10, with on-site tank storage used in Categories 11 and 12. Detention storage is provided by a pressure vessel in Categories 1 and 2, a looped underground pipeline in Categories 3 and 4, and a chlorine contact basin in Categories 5 through 12.

TABLE A-3. ESTIMATED COSTS FOR GROUNDWATER DISINFECTION USING OZONE

Category	Plant Capacity, mgd	Average Flow, mgd	Capital Cost, $1,000	Operation and Maintenance Cost $1,000/yr	Operation and Maintenance Cost ¢/1,000 gal	Total Cost, ¢/1,000 gal
1	0.06	0.013	17.0	4.0	110.4	165.1
2	0.14	0.045	37.2	4.6	31.6	61.6
3	0.31	0.133	48.8	9.2	19.5	31.6
4	0.96	0.400	99.7	10.9	7.4	15.5
5	3.06	2.30	181	14.2	3.0	7.5
6	7.52	3.25	341	19.5	1.6	5.0
7	15.44	6.75	592	29.1	1.2	4.0
8	25.33	11.50	883	41.8	1.0	3.5
9	44.15	20.00	1,360	64.1	0.9	3.1
10	123.88	55.50	2,406	151	0.8	2.1
11	465	205	5,346	450	0.6	1.4
12	1,505	650	10,960	1,500	0.6	1.2

Process includes direct in-line ozone application at a dosage of 1.0 mg/L followed by 5-minute contact time.

TABLE A-4. ESTIMATED COSTS FOR GROUNDWATER DISINFECTION AND STORAGE USING CHLORINE DIOXIDE

Category	Plant Capacity, mgd	Average Flow, mgd	Capital Cost, $1,000		O&M Cost, $1,000		Unit Cost, ¢/1,000 gal		
			Disinf.	Storage	Disinf.	Storage	Disinf.	Storage	Total
1	0.06	0.013	79.6	5.5	5.9	0.3	321.3	19.7	341.0
2	0.14	0.045	82.8	9.3	6.2	0.5	96.9	10.0	106.9
3	0.31	0.133	90	14	11.4	0.6	45.3	4.7	50.0
4	0.96	0.40	105	34	12.2	0.8	16.8	3.3	20.1
5	3.06	1.30	130	146	13.4	0	6.0	3.7	9.7
6	7.52	3.25	179	240	15.8	0	3.1	2.4	5.5
7	15.4	6.75	260	336	20.4	0	2.1	1.6	3.7
8	25.3	11.5	361	457	26.1	0	1.6	1.3	2.9
9	44.2	20.0	561	677	36.1	0	1.4	1.1	2.5
10	124	55.5	951	1,349	73.7	0	0.9	0.8	1.7
11	465	205	1,920	3,660	197	0	0.6	0.5	1.1
12	1,505	650	3,440	9,120	509	0	0.4	0.4	0.8

Costs are for a chlorine dioxide dose of 2.0 mg/l with a 15-minute contact time. Detention is in pressure vessels in Categories 1 and 2, in looped underground pipelines in Categories 3 and 4, and in chlorine contact basins in Categories 5 through 12.

TABLE A-5. ESTIMATED COSTS FOR GROUNDWATER DISINFECTION AND STORAGE USING CHLORAMINES

Category	Plant Capacity, mgd	Average Flow, mgd	Capital Cost, $1,000		O&M Cost, $1,000		Unit Cost, ¢/1,000 gal		
			Disinf.	Storage	Disinf.	Storage	Disinf.	Storage	Total
1	0.06	0.013	14	9	3.9	0.4	116.8	30.2	147.0
2	0.14	0.045	21	22	4.4	0.6	41.8	19.4	61.2
3	0.31	0.133	26	30	8.4	0.7	23.6	8.7	32.3
4	0.96	0.40	39	68	9.4	1.1	9.5	6.2	15.7
5	3.06	1.30	55	216	12.2	0	3.9	5.4	9.3
6	7.52	3.25	75	330	14.3	0	1.9	3.3	5.2
7	15.4	6.75	104	521	18.2	0	1.2	2.5	3.7
8	25.3	11.5	142	732	23.2	0	0.9	2.1	3.0
9	44.2	20.0	200	1,080	32.1	0	0.8	1.7	2.5
10	124	55.5	390	2,260	68.4	0	0.6	1.3	1.9
11	465	205	690	6,230	167	0	0.3	1.0	1.3
12	1,505	650	1,200	15,480	480	0	0.2	0.8	1.0

Costs are for a chlorine dose of 1.5 mg/l, an ammonia dose of 0.5 mg/l, and 30-minute detention. Chlorine is provided as a hypochlorite solution in Categories 1 through 4, by cylinder storage and feed in Categories 5 through 10, and is stored in on-site tanks in Categories 11 and 12. Ammonia is fed as anhydrous ammonia in Categories 1 through 4, and as aqua ammonia in Categories 5 through 12. Detention is in pressure vessels in Categories 1 and 2, in looped underground pipelines in Categories 3 and 4, and in chlorine contact chambers in Categories 5 through 12.

TABLE A-6. ESTIMATED COSTS FOR GROUNDWATER DISINFECTION USING ULTRAVIOLET LIGHT

Category	Plant Capacity, mgd	Average Flow, mgd	Capital Cost, $1,000	Operation and Maintenance Cost		
				$1,000/yr	¢/1,000 gal	¢/1,000 gal
1	0.06	0.013	11.4	0.9	15.2	61.9
2	0.14	0.045	15.4	1.1	7.8	20.2
3	0.31	0.133	23.7	2.0	4.3	10.1
4	0.96	0.400	59.9	4.1	2.8	7.6

Costs are for a dosage of 30,000 $\mu W \cdot sec/cm_2$ at a wavelength of 253.7 mm.

Appendix B: Surface Water Filtration Cost Calculations

Estimates of construction, capital, and operation and maintenance costs in Section VI are based on both the unit costs presented in that section and a set of design criteria. The design criteria, shown in Table B-1, represent commonly accepted or state-of-the-art practice for sizing chemical units and individual process structures. These criteria are also based on the assumption of a raw water within the normal ranges of water quality.

EXAMPLE CALCULATIONS

Most of the cost calculations in Section VI were made using computer programs developed by Culp/Wesner/Culp (now CWC-HDR, Inc.). The Water Cost Program (1986) was used predominantly, but calculations for a few processes (including slow-sand filters) were made using an in-house program of costs for small systems. Calculator computations were made for: (1) pressure vessels and looped pipelines used as detention facilities for small disinfection systems, (2) landfill user charges, and (3) process sizes above or below the range of input data of the computer programs.

Tables B-2 through B-7 present examples of input data and results from cost calculations for six different filtration process groups. Calculations for all the processes and factors in these groups, with the few exceptions noted above, were performed by the Water Cost Program.

It should also be noted that the Package Complete Treatment Plant in the Water Cost Program assumes a loading rate of 2 gpm/sf instead of the 5 gpm/sf cited in the design criteria in Table B-1. Therefore, for these calculations the design and average flows were modified to provide a cost for the same filter area that would be used by the actual flow at 5 gpm/sf. Other such adjustments and engineering judgements were made as necessary.

Costs in Tables B-2 and B-3 also contain supplemental storage, beyond the design criteria for clearwell storage, in order to make up the difference between plant design capacity and system design capacity, as described in Section VI.

Sludge Volumes

Sludge volumes for handling and disposal in the cost calculations were developed by sludge balance calculation based on several factors. The factors involved include: 1) assumed raw water turbidity; 2) chemical addition per design criteria in Table B-1; 3) assumed sludge densities (percent) from sludge handling processes, based on water utility experience; and 4) dewatered sludge densities of 50 percent (see Table B-1) from either sludge dewatering lagoons or mechanical sludge dewatering equipment for hauling to landfill.

Assumed raw water turbidities prior to filtration are 10 NTU for conventional treatment, conventional treatment with automatic backwashing filter, and complete treatment package plants. A raw water turbidity of 5 NTU was assumed for all forms of direct filtration processes, including diatomaceous earth filtration, and for slow-sand filtration.

An example of sludge balance calculations used, for the diatomaceous earth filtration plant in Table B-2, are as follows:

Total Solids Load

Assume:

a) 1 NTU = 1.5 mg/L suspended solids

b) D.E. precoat + body feed = 25 mg/L

Turbidity solids: 5 NTU x 1.5 mg/L/NTU x 8.34 lb/gal = 63 lb/MG

D.E. solids: 25 mg/L x 8.34 = 208 lb/MG

Total = 271 lb/MG

Effluent solids = 0.2 NTU x 1.5 x 8.34 = 3 lb/MG

Remaining solids for handling/disposal = 268 lb/MG

Say = 270 lb/MG

Sludge Dewatering Lagoon

1. Capacity (1 year of solids at design flow)

   For Category 1, design flow
   = 0.026 mgd

   Annual solids mass = (0.026 mgd)(270 lb/MG)(365 day/yr)
   = 2,560 lb/yr

   Assume solids concentration in D.E. filter backwash
   = 2.5 percent, then

   Annual solids volume (design)
   = (2,560 lb/yr)(1/8.34)(1/0.025)(1/7.5 gal/cu ft)
   = 1,640 cu ft
   Say = 1,700 cu ft/yr

   Provide two lagoons, one for filling while the other is drying
   = 3,400 cu ft/yr

2. Sludge removal for hauling

   Assume 1 yr actual (average) solids mass is removed, at 50 percent solids

   For Category 1, average flow
   = 0.013 mgd

$$\frac{(0.013 \text{ mgd})(270 \text{ lb/MG})(365)}{(8.34)(0.50)(7.5)} = 41 \text{ cu ft/yr}$$

Say = 50 cu ft/yr

TABLE B-1. PROCESS DESIGN CRITERIA

Chemical or Process	Design Criteria
Chemical Feed	
Alum	1) Conventional treatment - 20 mg/L 2) Direct filtration - 10 mg/L
Polymer	1) Conventional treatment - 0.3 mg/L 2) Direct filtration - 0.5 mg/L
Sodium Hydroxide or Lime	1) Conventional treatment - 10 mg/L 2) Direct filtration - 5 mg/L
Sulfuric Acid	21 gal/MG
Filtration Processes	
Rapid Mix	G = 900 sec^{-1}; 1-minute detention at design flow
Flocculation (Horizontal paddle)	G = 50 sec^{-1}; 30-minute detention at design flow (Q_D)
Rectangular Clarifier	Overflow rate = 1,000 gpd/sf at design flow; 2 basins minimum
Tube Settling Modules	2 gpm/sf rise rate at Q_D
Gravity Filtration	5 gpm/sf at Q_D
Convert Rapid-Sand Filters to Mixed-Media Filters	Surface area of rapid-sand filters
Filter-to-Waste Facilities	Sized to handle flow rate from one filter; not in Water Cost Program
Capping Sand Filters with Anthracite	Surface area of sand filter
Slow-Sand Filters	3 MG/acre/day (70 gpd/sf); not in Water Cost Program
Pressure Filtration	5 gpm/sf at Q_D
Contact Basins for Direct Filtration	60-minute detention at Q_D
Hydraulic Surface Wash	Filter area, sf
Backwash Rate	18 gpm/sf at Q_D

TABLE B-1 (Continued)

Chemical or Process	Design Criteria
Washwater Surge Basin	Store 20-minute volume of backwash flow from one filter at 18 gpm/sf
Automatic Backwashing Filter	Filter area, sf at 3 gpm/sf at Q_D
Clearwell Storage	Volume = 15% to 5% of Q_D for smaller to larger plants, respectively
Package Pressure Diatomite Filters	Design flow = Q_D; DE added at 15 mg/L
Pressure Diatomite Filters	Same as above
Package Ultrafiltration Plants	Design flow = Q_D
Package Conventional Complete Treatment	Design flow = Q_D in gpm; alum at 20 mg/L; polymer at 0.3 mg/L; soda ash at 10 mg/L
Disinfection Processes	
See Tables VI-12 through VI-15 for all processes except ultraviolet light, which is defined by plant flow in the Water Cost Program.	
Solids Handling Processes	
Sludge Holding Tanks	Volume = 2-day backwash volume; mixing energy = 0.25 hp/1,000 cf
Sludge Dewatering Lagoons	Volume = 1-year storage of sludge solids at 1%; hauling at 50% solids (based on average flow)
Liquid Sludge Hauling	Volume = 1 year of sludge at 5.0% solids (based on average flow)
Gravity Sludge Thickeners	Area = 150 gpd/sf sludge at 0.5%
Filter Press	Volume (cf) based on load of 5 lbs (dry) solids/cf/hr assuming 19 hrs/day operation; lime = 25% of dry solids
Dewatered Sludge Hauling	Volume (cy/yr) at 50% solids
Landfill Charges	Not in Water Cost Program; assumed $40/cy of dewatered sludge or $40/ton of liquid sludge

TABLE B-2. EXAMPLE CALCULATIONS FOR A 26,000 GPD DIRECT FILTRATION PLANT USING DIATOMACEOUS EARTH

Process or Factor	Design Parameter	Construction Cost	Operating Parameter	O&M Cost, $/yr	Unit Cost, ¢/1,000 gallons
Package Pressure Diatomite Filters	26,000 gpd	$ 67,900	13,000 gpd (0.5 T/yr DE)	$5,860	
Clearwell Storage Ground Level	46,400 gal	62,600	46,400 gal	0	
Sludge Dewatering Lagoons	3,400 cf	4,300	50 cf/yr removed	90	
Landfill User Charges		0	50 cf/yr	74	
SUBTOTAL, CONSTRUCTION		$134,800			
TOTAL, O&M				$6,024	127.0
Sitework, Interface Piping		$ 20,200			
Subsurface Consideration		6,700			
General Contractor's OH&P		19,400			
Engineering		27,200			
Legal, Fiscal, Administrative		6,900			
Interest During Construction		5,400			
TOTAL CAPITAL COST		$220,600			546.0
TOTAL, ¢/1,000 GALLONS					672.9

TABLE B-3. EXAMPLE CALCULATIONS FOR A 166,000 GPD COMPLETE TREATMENT PACKAGE PLANT

Process or Factor	Design Parameter	Construction Cost, $1,000	Operating Parameter	O&M Cost, $1,000/yr	Unit Cost, ¢/1,000 gallons
Package Complete Treatment Plant	46.3 gpm (@ 2 gpm/sf)	154.4	37.0 gpm (2 gpm/sf) 10.5 T/yr alum 310.25 lb/yr polymer 5.18 T/yr soda ash	39.8	
Clearwell Storage Washwater Storage Tanks	198,400 gal	98.7	198,400 gal	0	
Sludge Dewatering Lagoons	18,400 cf	7.2	1,470 cf/yr removed	0.4	
Landfill Charges		0	1,470 cf/yr	2.2	
SUBTOTAL, CONSTRUCTION		260.3			
TOTAL, O&M				42.4	89.2
Sitework, Interface Piping		39.0			
Subsurface Considerations		13.0			
General Contractor's OH&P		37.5			
Engineering		52.5			
Legal, Fiscal, Administrative		10.1			
Interest During Construction		15.2			
TOTAL CAPITAL COST		427.6			105.9
TOTAL, ¢/1,000 GALLONS					195.1

TABLE B-4. EXAMPLE CALCULATIONS FOR A 2.5 MGD DIRECT FILTRATION PLANT USING PRESSURE FILTERS

Process or Factor	Design Parameter	Construction Cost, $1,000	Operating Parameter	O&M Cost, $1,000/yr	Unit Cost, ¢/1,000 gallons
Alum Feed - Dry Stock	8.7 lb/hr	27.8	4.5 lb/hr	4.1	
Polymer Feed	10.4 lb/day	45.1	5.4 lb/day	6.2	
NaOH Feed	104.0 lb/day	23.7	54.0 lb/day	5.5	
Pressure Filtration Plant	360 sf	451.6	360 sf	48.8	
Filtration Media-Dual Media	360 sf	19.0	360 sf	0	
Backwash Pumping	1,620 gpm	70.7	360 sf	4.3	
Hydraulic Surface Wash	360 sf	53.5	360 sf	2.1	
Washwater Surge Basins	32,400 gal	164.9	32,400 gal	0	
Clearwell Storage - Ground	250,000 gal	213.8	250,000 gal	0	
Sludge Pumping-Unthickened	8.0 gal	41.1	4.5 gal	5.3	
Sludge Dewatering Lagoon	286,000 cf	69.5	152,000 cf/yr	2.1	
Landfill User Charges	-	-	152,000 cf/yr	22.5	
Administration Building	-	96.5	-	30.2	
SUBTOTAL, CONSTRUCTION		1,277.2			
TOTAL, O&M				131	27.6
Sitework, Interface Piping		177.1			
Subsurface Consideration		59.0			
General Contractor's OH&P		170.0			
Engineering		238.0			
Legal, Fiscal, Administrative		23.3			
Interest During Construction		135.7			
TOTAL CAPITAL COST		2,080.3			51.5
TOTAL, ¢/1,000 GALLONS					79.1

TABLE B-5. EXAMPLE CALCULATIONS FOR A 11.59 MGD CONVENTIONAL TREATMENT PLANT

Process or Factor	Design Parameter	Construction Cost, $1,000	Operating Parameter	O&M Cost, $1,000/yr	Unit Cost, ¢/1,000 gallons
Alum Feed - Liquid Stock	80.6 lb/hr	48.2	46.9 lb/hr	28.2	
Polymer Feed	29.0 lb/day	46.9	16.9 lb/day	8.4	
Lime Feed - No recalcn.	40.2 lb/hr	164.5	23.5 lb/hr	27.0	
Rapid Mix, G=900	1,080 cf	58.3	1,080 cf	38.2	
Flocculation Horizontal, G=50	32,300 cf	277.6	32,300 cf	8.7	
Rect. Clarifier, 6 units	1,950 sf(each)	1,150.8	1,950 sf(each)	33.8	
Gravity Filter Structure	1,640 sf	1,024.7	1,640 sf	56.9	
Filtration Media-Dual Media	1,640 sf	65.2	1,640 sf	0	
Backwash Pumping	7,380 gpm	145.6	1,640 sf	9.1	
Hydraulic Surface Wash	1,640 sf	123.6	1,640 sf	4.8	
Washwater Surge Basins	148,000 gal	464.4	148,000 gal	0	
Clearwell Storage - Below	936,000 gal	585.9	936,000 gal	0	
Sludge Dewatering Lagoons	1,324,000 cf	193.1	760,000 cf/yr	9.4	
Liquid Sludge Hauling, 20 miles	2.0 MG/yr	134.7	1.1 avg MG/yr	12.9	
In-Plant Pumping TDH = 50 feet	11.6 MGD	238.3	6.8 MGD	62.1	
Chem Sludge Pumping-Unthickened	50.0 GPM	54.4	25.0 GPM	6.1	
Landfill User Charges	-	-	1.1 MG/yr	112.6	
Administration Building	-	259.0	-	56.5	
SUBTOTAL, CONSTRUCTION		5,035.2			
TOTAL, O&M				474.5	19.3

TABLE B-5. Continued

Process or Factor	Design Parameter	Construction Cost, $1,000	Operating Parameter	O&M Cost, $1,000/yr	Unit Cost, ¢/1,000 gallons
Sitework, Interface Piping		716.4			
Subsurface Consideration		238.8			
General Contractor's OH&P		573.1			
Engineering		945.7			
Legal, Fiscal, Administrative		55.8			
Interest During Construction		689.2			
TOTAL CAPITAL COST		8,254.2			39.3
TOTAL, ¢/1,000 GALLONS					58.6

TABLE B-6. EXAMPLE CALCULATIONS FOR A 39.68 MGD DIRECT FILTRATION PLANT WITH FLOCCULATION PRECEDING GRAVITY FILTERS

Process or Factor	Design Parameter	Construction Cost, $1,000	Operating Parameter	O&M Cost, $1,000/yr	Unit Cost, ¢/1,000 gallons
Alum Feed - Liquid Stock	138 lb/hr	55.6	70.0 lb/hr	41.6	
Polymer Feed	165 lb/day	62.0	83.3 lb/day	20.8	
Lime Feed - No recalcn.	69.0 lb/hr	235.1	35.0 lb/hr	35.2	
Rapid Mix, G=900	3,690 cf	143.6	3,690 cf	115.6	
Flocculation Horizontal, G=50	110,700 cf	544.7	110,700 cf	22.9	
Gravity Filter Structure	5,580 sf	2,376.8	5,580 sf	143.7	
Dual Media	5,580 sf	167.2	5,580 sf	0	
Backwash Pumping	16,700 gpm	242.0	5,580 sf	20.5	
Hydraulic Surface Wash	5,580 sf	221.7	5,580 sf	11.7	
Washwater Surge Basins	335,000 gal	614.0	335,000 gal	0	
Clearwell Storage - Below	3,190,000 gal	1,243.7	3,190,000 gal	0	
Administration Bldg.	-	518.7	-	110.3	
In-Plant Pumping	39.7 mgd	579.7	20.0 mgd	166.4	
Thickened Sludge Pumps	25.0	20.1	15.0 gpm	4.2	
Unthickened Sludge Pumps	130.0 gpm	66.5	65.0 gpm	7.0	
Gravity Sludge Thickener - Lime, 2 Units	3,610 sf	632.2	3,610 sf	9.8	
Filter Press, Lime @ 90 T/yr	123 cf	1,929.2	123 cf	180.8	
Dewatered Sludge Hauling, 20 miles	4,370 cy/yr	116.2	2,850 cy/yr	9.1	
Landfill User Charges	-	-	2,850 cy/yr	114.0	
SUBTOTAL, CONSTRUCTION		9,769.0			
TOTAL, O&M				1,013.5	13.9

TABLE B-6. Continued

Process or Factor	Design Parameter	Construction Cost, $1,000	Operating Parameter	O&M Cost, $1,000/yr	Unit Cost, ¢/1,000 gallons
Sitework, Interface Piping		1,387.5			
Subsurface Consideration		462.5			
General Contractor's OH&P		999.0			
Engineering		1,814.9			
Legal, Fiscal, Administrative		85.0			
Interest During Construction		1,624.6			
TOTAL CAPITAL COST		16,142.5			26.0
TOTAL, ¢/1,000 GALLONS					39.9

TABLE B-7. EXAMPLE CALCULATIONS FOR A 404 MGD DIRECT FILTRATION PLANT WITH GRAVITY FILTERS AND CONTACT BASINS

Process or Factor	Design Parameter	Construction Cost, $1,000	Operating Parameter	O&M Cost, $1,000/yr	Unit Cost, ¢/1,000 gallons
Alum Feed - Liquid Stock	1,400 lb/hr	169.7	712 lb/hr	402.0	
Polymer Feed	1,683 lb/day	127.4	855 lb/day	164.1	
Lime Feed - With Recalcn.	702 lb/hr	304.9	356 lb/hr	132.2	
Gravity Filter Structure	57,600 sf	16,047.9	57,600 sf	1,205.4	
Dual Media	57,600 sf	1,573.7	57,600 sf	0	
Backwash Pumping	32,400 gpm	449.7	57,600 sf	137.6	
Hydraulic Surface Wash	57,600 sf	2,319.1	57,600 sf	101.3	
Washwater Surge Basins	648,000 gal	816.4	648,000 gal	0	
Contact Basin, 22 units	52,800 cf (ea)	3,891.8	52,800 cf (ea)	0	
Clearwell Storage - Below	20.2 MG	2,878.0	20.2 MG	0	
Administration Bldg.	-	1,748.6	-	334.4	
In-Plant Pumping	404 mgd	5,952.0	205 mgd	1,633.0	
Unthickened Sludge Pumps	1,300 gpm	136.4	670 gpm	18.9	
Thickened Sludge Pumps	260 gpm	55.3	135 gpm	13.2	
Gravity Sludge Thickener - Lime, 7 Units	17,000 sf	4,756.5	17,000 sf	85.3	
Filter Press, Lime @ 916 T/hr	1,250 cf	8,739.1	1,250 cf	1,034.5	
Dewatered Sludge Hauling, 20 miles	26,800 cy/yr	217.5	13,600 cy/yr	28.6	
Landfill User Charges	-	-	13,600 cy/yr	544.0	
SUBTOTAL, CONSTRUCTION		50,184.0			
TOTAL, O&M				5,835	7.8

TABLE B-7. Continued

Process or Factor	Design Parameter	Construction Cost, $1,000	Operating Parameter	O&M Cost, $1,000/yr	Unit Cost, ¢/1,000 gallons
Sitework, Interface Piping		7,265.3			
Subsurface Considerations		2,421.8			
General Contractor's OH&P		4,940.4			
Engineering		9,459.4			
Legal, Fiscal, Administrative		173.6			
Interest During Construction		12,580.5			
TOTAL CAPITAL COST		87,025.0			13.7
TOTAL, ¢/1,000 GALLONS					21.5

Appendix C: Costs of Obtaining an Exception to the Surface Water Filtration Rule

Under the new surface water filtration rule of the Safe Drinking Water Act (SDWA), a water supply utility may not have to filter its supply before distribution if it meets certain criteria. Each of the criteria is intended as a component of a complete system to safeguard public health. By continually meeting these criteria, a water utility replaces filtration as a barrier against microbial contamination with vigilant monitoring and reinforced disinfection capabilities.

Disinfection facilities for a utility meeting the exception criteria must: a) be capable of reducing Giardia cysts by 99.9 percent, and viruses by 99.99 percent; and b) have fully redundant capacity to make these reductions. Many water systems will have to add redundant facilities or increase the capacity of their disinfection system to meet this criterion, and some will have to do both. Representative costs for alternative approaches to reaching the required redundant disinfection capacity are shown in Tables C-1 through C-9. Other alternatives than those shown are possible, and conditions at a specific site are likely to differ from the costs here, but the tables do represent the relative magnitude of costs to achieve different levels of disinfection capacity (C·T products) and redundancy.

In addition to augmenting and upgrading disinfection facilities, a water utility will also have to implement rigorous monitoring programs. Some existing utilities already carry out some of the required procedures, but most will have to add these activities. Included in the complete program are:

- Annual sanitary surveys of a watershed and a complete and continuing watershed monitoring program.

- Add sampling and analysis for fecal coliforms to existing raw water monitoring programs.

- Add continuous turbidity monitoring of raw water supplies.

- Add capability to monitor for pH, temperature and chlorine residual following disinfection.

Detailed descriptions of these exception criteria procedures are beyond the scope of this document. They are, however, presented in the Guidance Document regarding new regulations of the SDWA.

Estimated costs of carrying out these exception criteria procedures are presented in Tables C-10 through C-14.

For purposes of consistency in comparisons with all other costs in this document, all annualized capital costs were obtained using an interest rate of 10 percent and a 20-year amorization period.

TABLE C-1. ESTIMATED COSTS FOR IMPROVING DISINFECTION BY ADDING REDUNDANT CHLORINE FEED FACILITIES

Category	Plant Capacity, mgd	Average Flow, mgd	Capital Cost, $1,000	Operation and Maintenance Cost		Total Cost, ¢/1,000 gal
				$1,000/yr	¢/1,000 gal	
1	0.026	0.013	1.5	0.2	4.2	7.9
2	0.068	0.045	3.0	0.3	1.8	3.9
3	0.166	0.133	4.4	0.6	1.2	2.3
4	0.50	0.40	7.5	1.1	0.8	1.4
5	2.50	1.30	11.2	1.5	0.3	0.6
6	5.85	3.25	17.5	2.5	0.2	0.4
7	11.59	6.75	28.2	3.9	0.2	0.3
8	22.86	11.50	35.3	5.0	0.1	0.2
9	39.68	20.00	45.9	7.3	0.1	0.2
10	109.90	55.50	85.2	16.2	0.1	0.1
11	404	205	204	52.4	0.1	0.1
12	1,275	650	334	142	0.1	0.1

Costs include redundant storage and feed facilities for an application rate of 5.0 mg/L plus auxiliary power (generator). These facilities and costs are considered to be a regulatory requirement for systems not already having redundant chlorine disinfection, but which are achieving greater than 99.9 percent inactivation of *Giardia* cysts with their existing disinfection facilities.

TABLE C-2. ESTIMATED COSTS FOR IMPROVING DISINFECTION BY ADDING OZONE

Category	Plant Capacity, mgd	Average Flow, mgd	Capital Cost, $1,000	Operation and Maintenance Cost $1,000/yr	Operation and Maintenance Cost ¢/1,000 gal	Total Cost, ¢/1,000 gal
1	0.026	0.013	19.0	4.8	101	148
2	0.068	0.045	35.6	5.5	30.4	55.9
3	0.166	0.133	62.8	11.0	23.4	38.6
4	0.50	0.40	128	13.1	8.9	19.2
5	2.50	1.30	317	17.0	3.6	7.8
6	5.85	3.25	559	24.2	2.0	7.5
7	11.59	6.75	964	32.5	1.3	4.6
8	22.86	11.50	1,258	40.3	1.0	4.5
9	39.68	20.00	1,800	61.3	0.8	3.7
10	109.90	55.50	3,690	146	0.7	2.8
11	404	205	9,800	538	0.7	2.2
12	1,275	650	28,460	1,420	0.6	2.0

Costs include ozone generation, feed, and contact (5 min) facilities for an application rate of 1.0 mg/L, plus redundant generation, feed, and auxiliary power facilities. These facilities will achieve greater than 99.9 percent inactivation of Giardia and 99.99 percent inactivation of viruses. These calculations also assume a demand of 0.5 mg/l after 5 minutes, for an effective CT of 3.0 at 0.5°C.

TABLE C-3. ESTIMATED COSTS FOR IMPROVING DISINFECTION BY ADDING NEW FACILITIES FOR OZONE AND CHLORINE AS A RESIDUAL

Category	Plant Capacity, mgd	Average Flow, mgd	Capital Cost, $1,000	Operation and Maintenance Cost		Total Cost, ¢/1,000 gal
				$1,000/yr	¢/1,000 gal	
1	0.026	0.013	23.4	6.7	141	210
2	0.068	0.045	44.4	7.6	46.3	78.1
3	0.166	0.133	76.0	15.3	31.5	49.9
4	0.50	0.40	153	21.3	14.6	26.9
5	2.50	1.30	350	28.7	6.0	14.7
6	5.85	3.25	610	43.4	3.7	9.7
7	11.59	6.75	1,050	62.9	2.6	7.6
8	22.86	11.50	1,360	79.2	1.9	5.7
9	39.68	20.00	1,940	117	1.6	4.7
10	109.90	55.50	3,940	270	1.3	3.6
11	404	205	10,390	940	1.2	2.8
12	1,275	650	29,400	2,500	1.1	2.6

Costs include chlorine storage and feed facilities for an application rate of 2.0 mg/L, and ozone generation, feed, and 5-min contact facilities for an application rate of 1.0 mg/L, plus redundant components for both ozone and chlorine facilities, and auxiliary power (generator). These facilities will achieve greater than 99.9 percent inactivation of *Giardia* and 99.99 percent inactivation of viruses, assuming an ozone demand of 0.5 mg/l after 5 minutes, i.e., an effective CT of 3.0 at 0.5°C.

TABLE C-4. ESTIMATED COSTS FOR IMPROVING DISINFECTION BY ADDING OZONE AND REDUNDANT Cl_2 TO EXISTING FACILITIES

Category	Plant Capacity, mgd	Average Flow, mgd	Capital Cost, $1,000	Operation and Maintenance Cost $1,000/yr	Operation and Maintenance Cost ¢/1,000 gal	Total Cost, ¢/1,000 gal
1	0.026	0.013	20.5	5.0	105	156
2	0.068	0.045	38.6	5.8	35.3	62.9
3	0.166	0.133	67.2	11.6	23.9	40.2
4	0.50	0.40	135	14.2	9.7	20.6
5	2.50	1.30	328	18.5	3.9	12.0
6	5.85	3.25	576	26.7	2.3	8.0
7	11.59	6.75	992	36.4	1.5	6.2
8	22.86	11.50	1,293	45.3	1.1	4.7
9	39.68	20.00	1,846	68.6	0.9	3.9
10	109.90	55.50	3,770	162	0.8	3.0
11	404	205	10,000	590	0.8	2.4
12	1,275	650	28,800	1,560	0.7	2.1

Costs include complete ozone generation, 5-min contact, and feed facilities for an application rate of 1.0 mg/L, plus redundant ozone and chlorine facilities, and auxiliary power (generator). These facilities will achieve greater than 99.9 percent inactivation of Giardia and 99.99 percent inactivation of viruses, assuming an ozone demand of 0.5 mg/l after 5 minutes, i.e., an effective C•T of 3.0 at 0.5°C.

TABLE C-5. ESTIMATED COSTS FOR IMPROVING DISINFECTION BY CONSTRUCTING NEW CHLORINE FACILITIES WITH REDUNDANCY AT C·T=100

Category	Plant Capacity, mgd	Average Flow, mgd	Capital Cost, $1,000	Operation and Maintenance Cost		Total Cost, ¢/1,000 gal
				$1,000/yr	¢/1,000 gal	
1	0.026	0.013	5.7	1.9	40.0	54.1
2	0.068	0.045	11.2	2.0	12.2	20.2
3	0.166	0.133	16.6	4.4	9.1	13.1
4	0.50	0.40	27.9	8.1	5.5	7.7
5	2.50	1.30	41.7	11.3	2.4	3.4
6	5.85	3.25	64.8	18.2	1.5	2.1
7	11.59	6.75	104	28.4	1.2	1.7
8	22.86	11.50	130	35.8	0.9	1.3
9	39.68	20.00	169	51.1	0.7	1.0
10	109.90	55.50	312	112	0.6	0.8
11	404	205	727	351	0.5	0.6
12	1,275	650	1,214	929	0.4	0.5

Based on separate new facilities with an application rate of 3.0 mg/L. These facilities are considered as a regulatory option for simulating a double barrier approach, i.e., requiring this as an on-line backup system to primary disinfection that achieves greater than 99.9 percent inactivation of *Giardia*.

TABLE C-6. ESTIMATED COSTS FOR IMPROVING DISINFECTION BY ADDING REDUNDANCY TO EXISTING FACILITIES AND INCREASING Cl_2 CAPACITY FROM C·T=100 to C·T=200

Category	Plant Capacity, mgd	Average Flow, mgd	Capital Cost, $1,000	Operation and Maintenance Cost		Total Cost, ¢/1,000 gal
				$1,000/yr	¢/1,000 gal	
1	0.026	0.013	5.7	0.4	8.0	22.1
2	0.068	0.045	10.6	0.7	4.1	11.7
3	0.166	0.133	15.0	1.3	2.7	6.3
4	0.50	0.40	23.7	2.8	1.9	3.8
5	2.50	1.30	33.3	6.7	1.4	2.3
6	5.85	3.25	48.6	12.7	1.1	1.6
7	11.59	6.75	77.0	19.8	0.8	1.1
8	22.86	11.50	94.9	30.4	0.7	1.0
9	39.68	20.00	122	46.0	0.7	0.9
10	109.90	55.50	222	104	0.6	0.7
11	404	205	509	333	0.4	0.6
12	1,275	650	790	883	0.3	0.4

Costs are based on existing storage and feed facilities for an application rate of 3.0 mg/L. These facilities are considered as a regulatory option for achieving greater than 99 percent inactivation of *Giardia* cysts. Assumes system already has adequate contact time and water quality characteristics that allow it to achieve C·T = 100 at an application rate of 3.0 mg/L, as determined by measuring residual (C, mg/L) near first customer. At low pH conditions and moderately low temperatures (e.g., 10°C), this may be adequate to achieve greater than 99.9 percent inactivation.

TABLE C-7. ESTIMATED COSTS FOR IMPROVING DISINFECTION BY ADDING REDUNDANCY TO EXISTING Cl_2 SYSTEM AND INCREASING CAPACITY 33 PERCENT [C·T FROM 150 TO 200]

Category	Plant Capacity, mgd	Average Flow, mgd	Capital Cost, $1,000	Operation and Maintenance Cost		Total Cost, ¢/1,000 gal
				$1,000/yr	¢/1,000 gal	
1	0.026	0.013	3.0	0.4	8.4	15.8
2	0.068	0.045	6.0	0.6	3.6	7.8
3	0.166	0.133	8.9	1.3	2.6	4.8
4	0.50	0.40	15.1	2.7	1.9	3.1
5	2.50	1.30	22.1	3.9	0.8	1.4
6	5.85	3.25	34.5	6.7	0.6	1.0
7	11.59	6.75	54.4	11.0	0.5	0.7
8	22.86	11.50	68.2	14.6	0.3	0.5
9	39.68	20.00	88.7	21.9	0.3	0.5
10	109.90	55.50	161	56.7	0.3	0.3
11	404	205	377	209	0.3	0.3
12	1,275	650	605	569	0.3	0.3

Cost are based on existing storage and feed facilities for an application rate of 5.0 mg/L. These facilities are considered as a regulatory option for achieving greater than 99 percent inactivation of *Giardia* cysts. Assumes system already has adequate contact time and water quality characteristics that allow it to achieve C·T = 100 at an application rate of 5.0 mg/L, as determined by measuring residual (C, mg/L) near first customer. At low pH conditions and moderately low temperatures (e.g., 10°C), this may be adequate to achieve greater than 99.9 percent inactivation.

TABLE C-8. ESTIMATED COSTS FOR IMPROVING DISINFECTION BY DOUBLING Cl_2 CAPACITY [C·T FROM 150 TO 300] AND ADDING REDUNDANCY FOR EXISTING FACILITIES

Category	Plant Capacity, mgd	Average Flow, mgd	Capital Cost, $1,000	Operation and Maintenance Cost		Total Cost, ¢/1,000 gal
				$1,000/yr	¢/1,000 gal	
1	0.026	0.013	6.1	0.4	8.5	23.6
2	0.068	0.045	11.4	0.8	4.9	13.0
3	0.166	0.133	16.0	1.7	3.5	7.4
4	0.50	0.40	25.6	3.7	2.6	4.7
5	2.50	1.30	36.2	8.7	1.8	2.7
6	5.85	3.25	52.9	17.8	1.5	2.1
7	11.59	6.75	84.1	27.7	1.2	1.6
8	22.86	11.50	104	42.8	1.0	1.3
9	39.68	20.00	133	65.6	0.9	1.1
10	109.90	55.50	244	151	0.8	0.9
11	404	205	575	469	0.7	0.8
12	1,275	650	875	1,278	0.6	0.6

Costs are based on existing storage and feed facilities for an application rate of 5.0 mg/L. These facilities are considered as a regulatory option for achieving greater than 99.9 percent inactivation of *Giardia* cysts.

TABLE C-9. ESTIMATED COSTS FOR IMPROVING DISINFECTION AND BY ADDING REDUNDANCY TO EXISTING Cl_2 SYSTEM INCREASING CAPACITY 20 PERCENT [C·T FROM 250 TO 300]

Category	Plant Capacity, mgd	Average Flow, mgd	Capital Cost, $1,000	Operation and Maintenance Cost		Total Cost, ¢/1,000 gal
				$1,000/yr	¢/1,000 gal	
1	0.026	0.013	1.8	0.3	6.3	10.7
2	0.068	0.045	3.7	0.4	2.4	5.0
3	0.166	0.133	5.8	0.8	1.6	3.0
4	0.50	0.40	10.5	1.5	1.1	1.9
5	2.50	1.30	16.2	2.3	0.5	0.9
6	5.85	3.25	26.0	4.0	0.3	0.6
7	11.59	6.75	43.0	6.7	0.3	0.5
8	22.86	11.50	55.3	9.0	0.2	0.4
9	39.68	20.00	75.9	14.6	0.2	0.3
10	109.90	55.50	147	37.3	0.2	0.2
11	404	205	303	141	0.2	0.2
12	1,275	650	537	427	0.2	0.2

Cost are based on existing storage and feed facilities for an application rate of 5.0 mg/L. These facilities are considered as a regulatory option for achieving greater than 99.9 percent inactivation of *Giardia* cysts.

TABLE C-10. SANITARY SURVEY AND WATERSHED MANAGEMENT PROGRAM

Category	Average Flow, mgd	Annual Cost, $1,000/yr	Annual Cost, ¢/1,000 gal
1	0.013	1.0	21.1
2	0.045	2.4	14.6
3	0.133	4.6	9.5
4	0.40	8.7	6.0
5	1.30	17.2	3.6
6	3.25	29.4	2.5
7	6.75	45.0	1.8
8	11.50	61.4	1.5
9	20.00	84.8	1.2
10	55.50	154	0 8
11	205	329	0.4
12	650	644	0.3

Notes: 1. Costs are incremental values estimated as necessary to upgrade existing programs to levels adequate to meet requirements of the new regulations.

2. A value of $1,000/yr was assumed for Category 1. This value assumes system already has good program, e.g., needs only an annual survey. Costs greater than this might be prohibitive for such a small system.

3. A value of $325,000/yr was used to represent systems in Category 11, based on current estimates of incremental costs seen as necessary to meet program requirements by the Seattle Water Department. It is assumed that other systems of similar size would have to incur similar costs.

4. Other costs (including the final Category 11 costs) were estimated from the following exponential equation, derived from the two values given above:

$Y = 14.8(x)^{0.583}$
Y = $1,000/yr
X = Average flow, mgd

TABLE C-11. RAW WATER COLIFORM MONITORING COSTS

Category	Average Flow, mgd	Annual Cost, $1,000/yr	Annual Cost, ¢/1,000 gal
1	0.013	1.5	31.6
2	0.045	1.5	9.1
3	0.133	3.0	6.2
4	0.40	3.0	2.1
5	1.30	4.5	1.0
6	3.25	6.0	0.5
7	6.75	3.1	0.1
8	11.50	3.1	0.1
9	20.00	3.1	0.0
10	55.50	3.1	0.0
11	205	3.1	0.0
12	650	3.1	0.0

Note: Costs for Categories 1-6 assume analyses performed by private laboratory at $29 per sample. Costs for Categories 7-12 assume analyses performed by water utility personnel and equipment at $12 per sample. Assumed sample frequencies are 1/week for Categories 1 and 2, 2/week for Categories 3 and 4, 3/week for Category 5, 4/week for Category 6, and 5/week for Categories 7 through 12. Lower costs might be incurred if total coliform sampling and analysis are used instead of fecal coliform monitoring.

TABLE C-12. TURBIDITY MONITORING COSTS

Category	Average Flow, mgd	Capital Cost, $1,000	Annualized Capital Cost, $/yr	Annual Cost, ¢/1,000 gal
1	0.013	2.0	235	5.0
2	0.045	2.0	235	1.4
3	0.133	2.0	235	0.5
4	0.40	2.0	235	0.2
5	1.30	2.0	235	0.0
6	3.25	2.0	235	0.0

Note: Assumes purchase of a continuous turbidimeter and two strip chart recorders for Categories 1-6. Costs for systems in Categories 7-12 are considered negligible as part of their total operating costs and are therefore not considered here.

TABLE C-13. COSTS FOR pH, TEMPERATURE, AND CHLORINE RESIDUAL MONITORING EQUIPMENT

Category	Average Flow, mgd	Capital Cost, $1,000	Annualized Capital Cost, $/yr	Annual Cost, ¢/1,000 gal
1	0.013	5.5	646	13.6
2	0.045	5.5	646	3.9
3	0.133	5.5	646	1.3
4	0.40	5.5	646	0.4
5	1.30	5.5	646	0.1
6	3.25	5.5	646	0.1

Note: Equipment costs for Categories 1-6 are assumed to be a total of $5,200 for chlorine residual monitoring and $300 for a pH-temperature meter. Equipment requirements and capital costs are the same for Categories 7-12 as for Categories 1-6, but are negligible as a part of the total operation costs of larger systems, and are therefore not considered here.

TABLE C-14. SUMMARY OF ANNUAL COSTS TO MEET EXCEPTION CRITERIA

Category	Average Flow, mgd	Total Annual Cost, $1,000/yr	Total Annual Cost, ¢/1,000 gal
1	0.013	3.5	58.5
2	0.045	4.9	25.4
3	0.133	8.6	16.2
4	0.40	12.7	8.2
5	1.30	22.6	4.5
6	3.25	36.3	3.0
7	6.75	48.1	2.0
8	11.50	64.5	1.5
9	20.00	87.9	1.2
10	55.50	157	0.8
11	205	332	0.4
12	650	647	0.3

Notes: 1. The costs above are representative only of systems already achieving 99.9 percent inactivation of *Giardia* cysts. Costs for systems achieving less than that percent would be greater than the costs above, since they would also have to implement one or more of the stops shown in Tables C-1 through C-9.

2. Annual costs above are sums of values from Tables C-10 through C-13, e.g., for Category 5:

Item	Annual Cost, $1,000/yr
Sanitary survey and watershed management program	17.2
Raw water coliform monitoring	4.5
Turbidity monitoring	0.24
pH, temperature, and chlorine residual monitoring	0.65
TOTAL ANNUAL COST	$22.6

Appendix D: Costs for Presently Filtering Systems to Improve Their Disinfection Facilities

Several alternatives for upgrading disinfection systems were developed utilizing plant-specific information provided in a survey conducted by the AWWA (1987). The disinfection performance of the plants in the AWWA survey was reviewed in comparison to disinfection levels recommended by EPA. The number of plants which do not maintain a) a 0.5 log inactivation of *Giardia* cysts b) a 1 log inactivation of *Giardia* cysts and c) a 2 log inactivation of *Giardia* cysts as well as effective inactivation of enteric viruses were determined. Based on this plant specific information, general design criteria and assumptions were developed as a basis for determining approximate cost estimates associated with the treatment alternatives considered for upgrading disinfection. Specific information reported in the survey about the physical layout of each of the plants was used as the basis for cost estimates. Thus, the costs presented here are not applicable to specific sites.

ALTERNATIVES EVALUATED

The alternatives evaluated for providing increased disinfection were:

- Increase chlorine or ozone dose
- Install additional storage for disinfectant contact volume
- Baffle clearwells
- Move the ammonia addition downstream of the chlorine addition for chloramine systems or move chlorine addition upstream for chlorine systems
- Apply ozone or chlorine dioxide as alternative disinfectants

Increased Dose

The alternative of increasing the disinfectant dose was evaluated. For systems in the AWWA survey using free chlorine and achieving less than a 2 log *Giardia* cyst inactivation, the average increase in dose needed to upgrade to achieve a 0.5 log, 1 log and 2 log *Giardia* inactivation was 0.3mg/L, 0.5 mg/L and 0.81 mg/L respectively. These doses were used to generate costs for the 12 USEPA flow categories presented in Tables D-1, D-2, and D-3.

Additional Contact Time

Another alternative was to increase disinfectant contact time by providing additional storage after the clearwell, without changing the chemical doses. This alternative is appropriate for utilities which maintain a high residual leaving the plant and will not require an inordinate amount of storage to provide the additional inactivation needed. For systems using free chlorine and achieving less than a 2 log *Giardia* cyst inactivation, the average increase in contact time needed to upgrade to achieve a 0.5 log, 1 log and 2 log *Giardia* cyst inactivation was determined to be 10 minutes, 43 minutes and 73 minutes respectively. In sizing the storage tanks, it was assumed that the contact time for disinfection is 40 percent of the theoretical detention time. The storage volume for the 12 USEPA flow categories was based on the design flow.

Tables D-1, D-2 and D-3 present costs for both increasing chlorine dose and adding contact storage, in order to meet three different levels of *Giardia* cyst inactivation.

Additional Baffles in Clearwell

The option of adding baffles was evaluated when the detention time of a clearwell was known. The detention time reported by AWWA was 40 percent of the theoretical detention time. This was the detention time used to calculate the CTs maintained in the plants. Installation of baffles in the clearwell was assumed to increase the actual detention time of the clearwell to 90 percent of the theoretical detention time. In developing costs for the USEPA flow categories, it was assumed that the clearwells have a volume equal to 10 percent of the design plant flow.

Capital costs for retrofitting clearwells with additional baffles are shown in Table D-4.

Moving the Point of Ammoniation/Chlorination

This alternative was developed for systems disinfecting with chloramines. The ammonia application is moved from the point of chlorination to a point further downstream to allow for free chlorine contact time to inactivate viruses and Giardia. Only an additional capital cost is incurred for this

TABLE D-1

COSTS TO PROVIDE 2 LOG GIARDIA CYST INACTIVATION BY INCREASING CHOLORINE DOSE OR CONTACT TIME

EPA Flow Capacity	Average Flow (mgd)	Increase Chlorine Dose* ($1,000/yr)	Plant Capacity (mgd)	Additional Storage** ($1,000)
1	0.013	0	0.026	46
2	0.045	0	0.068	58
3	0.133	0	0.166	78
4	0.40	0	0.50	120
5	1.3	1	2.50	260
6	3.25	1	5.85	440
7	6.75	2	11.59	700
8	11.50	5	22.89	1,100
9	20.00	8	39.68	1,600
10	55.50	22	109.9	4,000
11	205	83	404	15,000
12	650	263	1,275	40,000

Notes:

* Chlorine dose of 0.81 mg/l is assumed.
**Contact time of 73 minutes is assumed.

TABLE D-2

COSTS TO PROVIDE 1 LOG GIARDIA CYST INACTIVATION BY INCREASING CHLORINE DOSE OR CONTACT TIME

EPA Flow Capacity	Average Flow (mgd)	Increase Chlorine Dose* ($1,000/yr)	Plant Capacity (mgd)	Additional Storage** ($1,000)
1	0.013	0	0.026	42
2	0.045	0	0.068	52
3	0.133	0	0.166	67
4	0.40	0	0.50	100
5	1.3	0	2.50	220
6	3.25	0	5.85	340
7	6.75	1	11.59	520
8	11.50	2	22.89	820
9	20.00	4	39.68	1,250
10	55.50	12	109.9	2,900
11	205	44	404	9,400
12	650	139	1,275	28,000

Notes:

* Chlorine dose of 0.5 mg/l is assumed.
**Contact time of 43 minutes is assumed.

TABLE D-3

COSTS TO PROVIDE 0.5 LOG GIARDIA CYST INACTIVATION
BY INCREASING CHLORINE DOSE OR CONTACT TIME

EPA Flow Capacity	Average Flow (mgd)	Increase Chlorine Dose* ($1,000/yr)	Plant Capacity (mgd)	Additional Storage** ($1,000)
1	0.013	0	0.026	27
2	0.045	0	0.068	35
3	0.133	0	0.166	46
4	0.40	0	0.50	63
5	1.3	0	2.50	105
6	3.25	0	5.85	140
7	6.75	0	11.59	200
8	11.50	1	22.89	270
9	20.00	3	39.68	440
10	55.50	7	109.9	700
11	205	26	404	800
12	650	83	1,275	850

Notes:

* Chlorine dose of 0.3 mg/l is assumed.
**Contact time of 10 minutes is assumed.

TABLE D-4

COSTS FOR RETROFITTING EXISTING CLEARWELLS WITH BAFFLES*

EPA Flow Capacity	Plant Capacity (mgd)	Cost ($1,000)
1	0.026	1
2	0.068	1.5
3	0.166	2.5
4	0.50	4.8
5	2.50	16
6	5.85	35
7	11.59	68
8	22.89	140
9	39.68	280
10	109.9	850
11	404	3,600
12	1,275	13,000

Note:

* Clear well capacity assumed to be 10 percent of plant design flow.

alternative, since there is no operation and maintenance cost for piping. The capital cost includes the cost for piping to carry the ammonia solution to new points of application downstream of the next unit process. There may be several injection points off the main feed line to allow for a change in the ammonia application point with changing seasonal needs.

When more than one point of chloramine application was in use at the plant, the change in chemical application was made at the second or third application point, in order to limit trihalomethane (THM) formation, if there was sufficient detention time available in this section of the plant. It is assumed that small systems, in flow categories 1 through 4 do not use chloramines. Therefore, this is only applicable in flow categories 5 through 12.

Since the costs associated with moving the point of chemical application are for increased piping, these costs can also be applied to systems disinfecting with chlorine. The disinfectant contact time can be increased by moving the point of chlorine application to an upstream point in the plant. Capital costs for moving the point of ammoniation/chlorination are shown in Table D-5.

Ozone Disinfection

Ozonation may be practiced at many points in a plant. Ozonation prior to filtration was assumed. It was also assumed that a secondary disinfectant is already in place to provide a residual for the distribution system.

The remaining residual for a given dose of ozone cannot be accurately predicted in different waters because ozone demand varies greatly depending on water quality characteristics, such as pH, alkalinity, hardness, temperature, and organic content. Typical ozone doses range from 1 to 4 mg/L. Some uncommon cases require an ozone dose as high as 8 to 10 mg/L. Therefore, costs for the USEPA flow categories were developed for ozone doses of 1 to 4 mg/L, and a contact time of 10 minutes, as shown in Table D-6. The effective CT is assumed to achieve 0.5 to 2.0 log inactivation of Giardia cysts depending on water quality. The ozone contactors are baffled units with an even distribution of ozone application. The dose is a total dose applied in the contactor.

TABLE D-5

ESTIMATED COSTS FOR MOVING THE POINT OF AMMONIATION/CHLORINATION

EPA Flow Category	Capital Cost ($1,000)
1-4	25
5-6	35
7-9	60
10-11	85
12	200

TABLE D-6

COSTS FOR DISINFECTION WITH OZONE AT VARIOUS DOSES

Flow Category	Average Flow (mgd)	Plant Capacity (mgd)	1 mg/L Capital ($1,000)	1 mg/L O&M ($1,000/yr)	2 mg/L Capital ($1,000)	2 mg/L O&M ($1,000/yr)	3 mg/L Capital ($1,000)	3 mg/L O&M ($1,000/yr)	4 mg/L Capital ($1,0000)	4 mg/L O&M ($1,000/yr)
1	0.013	0.026	9	3.5	13	3.5	15	4	15	4
2	0.045	0.068	17	4	25	4	30	4.5	30	5
3	0.133	0.166	32	4.5	50	5	55	6	60	6
4	0.4	0.5	70	6	105	7	120	8	135	10
5	1.3	2.5	220	10	330	15	400	17	460	20
6	3.25	5.85	400	20	620	25	760	30	900	40
7	6.75	11.59	680	30	1,000	45	1,300	60	1,500	80
8	11.5	22.89	1,100	40	1,600	80	2,100	85	2,500	115
9	20	39.68	1,600	55	2,400	100	3,200	140	3,800	190
10	55.5	109.9	3,200	135	5,000	230	7,000	350	8,200	450
11	205	404	8,000	400	12,500	790	18,000	1,200	22,000	1,700
12	650	1,295	18,000	1,200	28,000	2,400	42,000	3,400	52,000	5,000

Chlorine Dioxide Disinfection

The addition of chlorine dioxide after the filtration process was also considered. With such an addition, any current application of another disinfectant at this point would be discontinued. A chlorine dioxide dose of 1.5 mg/L is assumed to produce less than 1.0 mg/L of chlorites, chlorates and chlorine dioxide in sum, which is the current EPA guideline. These conditions with existing clearwell storage are assumed to achieve 0.5 to 2 log inactivation of Giardia cysts depending on source water quality. It is also assumed that there is already a secondary disinfection point in place to provide a residual for the distribution system.

Table D-7 shows both capital and operating costs for installing and operating a chlorine dioxide system.

Combined Alternatives

In some of the utilities, a combination of alternatives would be appropriate to achieve the required level of Giardia cyst inactivation, while avoiding increased THM formation. The combined alternatives in each table include individual alternatives as follows:

Alternative 1. Move the point of ammoniation, increase chlorine dose, retrofit clearwells with baffles, and install additional contact storage.

Alternative 2. Move the point of ammoniation, increase chlorine dose and install ozone as an additional disinfectant.

For each of the combined alternatives it is assumed that two of the individual alternatives will be applied. Thus, the costs are determined as the average of the individual alternatives multiplied by two. However, for system flow categories 1 through 4, chloramines are not applied and the alternative of moving the point of ammoniation is not applicable. The costs used to generate the averages were the costs for each of the individual alternatives used independently to increase disinfection to the specified level. This results in a conservative cost estimate since the increase in chlorine dose, additional storage and ozone dose needed will be less when alternatives are combined. Also, an ozone dose of 4 mg/L was used as a conservative estimate.

Costs for the two possible combined alternatives are presented in Tables D-8, D-9, and D-10 for the specified levels of disinfection.

TABLE D-7

COSTS TO PROVIDE DISINFECTION BY CHLORINE DIOXIDE*

EPA Flow Capacity	Plant Capacity (mgd)	Capital ($1,000)	Average Flow (mgd)	O&M ($1,000/yr)
1	0.026	130	0.013	5.8
2	0.068	135	0.045	6.1
3	0.166	145	0.133	6.6
4	0.50	165	0.40	7.8
5	2.50	230	1.3	10
6	5.85	280	3.25	13
7	11.59	360	6.75	16
8	22.89	450	11.50	19
9	39.68	640	20.00	24
10	109.9	1,150	55.50	38
11	404	1,800	205	68
12	1,275	2,200	650	100

Note:

*Chlorine dioxide dose of 1.5 mg/l is assumed.

TABLE D-8

ESTIMATED COSTS FOR COMBINING ALTERNATIVES
2 LOG GIARDIA INACTIVATION

EPA Flow Category	Alternative 1* Capital ($1,000)	Alternative 1* O&M ($1,000/yr)	Alternative 2* Capital ($1,000)	Alternative 2* O&M ($1,000/yr)
1	30	0	16	4
2	40	0	30	6
3	40	0	60	6
4	80	0	140	10
5	160	0	330	14
6	260	0	620	28
7	410	2	1,000	54
8	650	2	1,700	80
9	1,000	4	2,600	130
10	2,500	12	5,500	310
11	9,300	42	14,700	1,200
12	26,600	130	34,800	3,500

Notes:

*Relocated ammonia feeders are installed downstream from chlorine inlet in order to provide a CT equal to 6. A chlorine dose of 0.8 mg/l is assumed. Costs to retrofit existing clearwells with baffles assume clearwell capacity to be 10 percent of plant design flow and includes material, labor, overhead and profit, and contingencies. Additional storage assumes a contact of 73 minutes.

**Relocated ammonia feeders are installed downstream from chlorine inlet in order to provide a CT equal to 6. A chlorine dose of 0.8 mg/l is assumed. Ozone, the alternate disinfectant, is dosed at 4 mg/l.

TABLE D-9

ESTIMATED COSTS FOR COMBINING ALTERNATIVES
1 LOG GIARDIA INACTIVATION

EPA Flow Category	Alternative 1* Capital ($1,000)	Alternative 1* O&M ($1,000/yr)	Alternative 2* Capital ($1,000)	Alternative 2* O&M ($1,000/yr)
1	28	0	16	4
2	36	0	30	6
3	44	0	60	6
4	70	0	140	10
5	140	0	330	14
6	200	0	620	28
7	320	0	1,000	54
8	510	2	1,700	78
9	800	2	2,600	130
10	1,920	6	5,500	310
11	6,540	22	14,700	1.160
12	20,600	70	34,800	3,430

Notes:

*Relocated ammonia feeders are installed downstream from chlorine inlet in order to provide a CT equal to 6. A chlorine dose of 0.8 mg/l is assumed. Costs to retrofit existing clearwells with baffles assume clearwell capacity to be 10 percent of plant design flow and includes material, labor, overhead and profit, and contingencies. Additional storage assumes a contact of 73 minutes.

**Relocated ammonia feeders are installed downstream from chlorine inlet in order to provide a CT equal to 6. A chlorine dose of 0.8 mg/l is assumed. Ozone, the alternate disinfectant, is dosed at 4 mg/l.

TABLE D-10

ESTIMATED COSTS FOR COMBINING ALTERNATIVES
0.5 LOG GIARDIA INACTIVATION

EPA Flow Category	Alternative 1* Capital ($1,000)	Alternative 1* O&M ($1,000/yr)	Alternative 2* Capital ($1,000)	Alternative 2* O&M ($1,000/yr)
1	20	0	16	4
2	25	0	30	6
3	30	0	60	6
4	50	0	140	10
5	80	0	330	14
6	110	0	620	30
7	160	0	1,000	50
8	240	0	1,700	80
9	400	2	2,600	130
10	800	4	5,500	300
11	2,200	14	14,700	1,200
12	7,000	14	34,800	3,400

Notes:

*Relocated ammonia feeders are installed downstream from chlorine inlet in order to provide a CT equal to 6. A chlorine dose of 0.8 mg/l is assumed. Costs to retrofit existing clearwells with baffles assume clearwell capacity to be 10 percent of plant design flow and includes material, labor, overhead and profit, and contingencies. Additional storage assumes a contact of 73 minutes.

**Relocated ammonia feeders are installed downstream from chlorine inlet in order to provide a CT equal to 6. A chlorine dose of 0.8 mg/l is assumed. Ozone, the alternate disinfectant, is dosed at 4 mg/l.

Appendix E: Cost Range Information Regarding Land, Piping, and Finished Water Pumping

Several cost components have been intentionally omitted from calculations presented in the preceding chapters and appendices. These components are ones which are highly variable depending on site-specific conditions. Included among these components are the cost of land on which treatment facilities might be built, the cost of a pipeline from a surface water source to a treatment site, and the cost of finished water pumping.

As supplemental information to cost estimates presented previously, the following tables provide ranges of costs for land, piping, and finished water pumping. Because of the potentially wide variation in these costs, the low and high values in these tables should be thought of as representative of high and low cost ranges, not as extremes.

TABLE E-1. ESTIMATED COST OF PURCHASING LAND FOR SURFACE WATER TREATMENT FACILITIES

Category	Plant Capacity, mgd	Average Flow, mgd	Land Required, acres	Land Cost, $1,000	Total Cost, ¢/1,000 gal
1	0.026	0.013	0.33	(1) 1.65	4.0
				(2) 16.5	41.0
				(3) 33.0	82.0
2	0.068	0.045	0.33	(1) 1.65	1.0
				(2) 16.5	12.0
				(3) 33	24.0
3	0.166	0.133	0.75	(1) 3.75	1.0
				(2) 37.5	9.0
				(3) 75	18.0
4	0.50	0.40	1.0	(1) 5	0.4
				(2) 50	4.0
				(3) 100	8.0
5	2.50	1.30	1.5	(1) 7.5	0.2
				(2) 75	2.0
				(3) 150	4.0
6	5.85	3.25	2.0	(1) 10	0.1
				(2) 100	1.0
				(3) 200	2.0
7	11.59	6.75	3.0	(1) 15	0.07
				(2) 150	0.7
				(3) 300	3.0
8	22.86	11.50	6.0	(1) 30	0.1
				(2) 300	0.8
				(3) 600	1.7
9	39.68	20.00	8.0	(1) 40	0.1
				(2) 400	0.6
				(3) 800	1.3
10	109.90	55.50	16.0	(1) 80	0.05
				(2) 800	0.5
				(3) 1,600	0.9
11	404	205	56	(1) 280	0.04
				(2) 2,800	0.4
				(3) 5,600	0.9
12	1,275	650	166	(1) 830	0.04
				(2) 8,300	0.4
				(3) 16,600	0.8

Notes:
Categories 1 through 5 assume use of package treatment plants.
Categories 6 through 12 assume use of conventional treatment plants.
Land cost assumed:
(1) Low: $5,000/acre
(2) Medium: $50,000/acre
(3) High: $100,000/acre

TABLE E-2. ESTIMATED COST OF PIPING FROM SURFACE WATER SOURCE TO FILTRATION PLANT

Category	Plant Capacity, mgd	Average Flow, mgd	Pipe Diam. inches	Capital Cost, $1,000	O&M $1,000/yr	O&M ¢/1,000 gal	Total Cost, ¢/1,000 gal
1	0.026	0.013	6	(1) 16.00	(1) 1.0	21	61.0
				(2) 163.00	(2) 2.0	42	442.0
				(3) 1,635.00	(3) 3.0	63	4,063.0
2	0.068	0.045	6	(1) 16.00	(1) 1.0	6	18.0
				(2) 163.00	(2) 2.0	12	129.0
				(3) 1,635.00	(3) 3.0	18	1,188.0
3	0.166	0.133	6	(1) 16.00	(1) 1.0	2	6.0
				(2) 163.00	(2) 2.0	4	44.0
				(3) 1,635.00	(3) 3.0	6	402.0
4	0.50	0.40	6	(1) 16.00	(1) 1.0	0.7	2.0
				(2) 163.00	(2) 2.0	1.4	14.4
				(3) 1,635.00	(3) 3.0	2.0	130.0
5	2.50	1.30	10	(1) 27.00	(1) 1.0	0.2	0.9
				(2) 272.00	(2) 2.0	0.4	7.1
				(3) 2,724.00	(3) 3.0	0.6	67.4
6	5.85	3.25	14	(1) 38.00	(1) 1.0	0.08	0.5
				(2) 381.00	(2) 2.0	0.2	4.0
				(3) 3,814.00	(3) 19.1	1.6	40.0
7	11.59	6.75	20	(1) 54.00	(1) 2.0	0.08	0.3
				(2) 545.00	(2) 3.0	0.1	3.0
				(3) 5,450.00	(3) 27.0	1.1	27.0
8	22.86	11.50	30	(1) 82.00	(1) 2.0	0.05	0.3
				(2) 817.00	(2) 4.0	0.1	2.4
				(3) 8,170.00	(3) 40.0	1.0	24.0
9	39.68	20.00	36	(1) 98.00	(1) 2.0	0.03	0.2
				(2) 981.00	(2) 5.0	0.07	1.7
				(3) 9,808.00	(3) 50.0	1.20	17.0
10	109.90	55.50	60	(1) 163.00	(1) 3.0	0.01	0.1
				(2) 1,635.00	(2) 8.0	0.04	1.0
				(3) 16,347.00	(3) 82.0	0.40	9.4
11	404	205	108	(1) 294.00	(1) 3.0	-	0.05
				(2) 2,940.00	(2) 15.0	0.02	0.5
				(3) 29,400.00	(3) 147.0	0.20	5.0
12	1,275	650	192	(1) 523.00	(1) 3.0	-	0.03
				(2) 5,230.00	(2) 26.0	0.01	0.30
				(3) 52,300.00	(3) 261.0	0.10	3.0

Notes:

1. Costs are for three pipe lengths: (1) 0.1 mile, (2) 1 mile, and (3) 10 miles.
2. Construction cost of pipe estimated at $3/in-diam/foot with added 72% for contingencies and other related costs. The capital cost of the pipe is therefore $5.16/in-diam/foot.
3. O&M costs are assumed at: (1) $1,000/year, (2) $2,000/year, and (3) $3,000/year for the low flows. At the higher flows, O&M costs are calculated at 0.5% of the capital cost with a minimum of $2,000/year and $3,000/year for the 0.1-mile pipe length.

TABLE E-3. ESTIMATED COSTS FOR FINISHED WATER PUMPING

Category	Plant Capacity, mgd	Average Flow, mgd	Capital Cost, $1,000		Operation & Maintenance Cost $1,000/year	¢/1,000 gals	Total Cost, ¢/1,000 gal
1	0.026	0.013	(1)	17.4	0.6	12.6	55.2
			(2)	22.6	0.7	14.3	70.3
			(3)	31.6	0.8	16.8	78.2
2	0.068	0.045	(1)	19	0.84	5.1	18.7
			(2)	24.6	1.0	6.1	23.7
			(3)	34.4	1.2	7.3	31.9
3	0.166	0.133	(1)	21	2.4	4.9	10.0
			(2)	27.0	2.9	6.0	12.7
			(3)	37.8	3.5	7.2	16.3
4	0.50	0.40	(1)	26	5.2	3.6	5.7
			(2)	33.6	6.3	4.3	7.0
			(3)	47	7.6	5.2	9.0
5	2.50	1.30	(1)	135	40.5	9.0	12.3
			(2)	175	48	10.1	14.4
			(3)	243	58.5	12.3	18.4
6	5.85	3.25	(1)	250	82	6.9	9.4
			(2)	325	103	8.7	11.9
			(3)	457	135	11.4	16.0
7	11.59	6.75	(1)	410	140	5.7	7.6
			(2)	520	185	7.5	10.0
			(3)	792	272	11.1	14.9
8	22.86	11.50	(1)	705	205	4.9	6.9
			(2)	930	285	6.8	9.4
			(3)	1,215	459	10.9	14.3
9	39.68	20.00	(1)	1,150	360	4.9	6.7
			(2)	1,550	510	7.0	9.5
			(3)	2,000	792	10.9	13.9
10	109.90	55.50	(1)	2,750	900	4.4	6.0
			(2)	3,700	1,300	6.5	8.6
			(3)	5,055	2,184	10.8	13.7
11	404	205	(1)	9,300	3,200	4.3	5.8
			(2)	12,500	5,000	6.7	8.7
			(3)	18,150	8,048	10.8	13.7
12	1,275	650	(1)	31,000	10,500	4.4	5.9
			(2)	44,000	16,000	6.7	8.9
			(3)	56,340	25,508	10.8	13.7

Notes:
Assume pumping pressures
(1) Low: 100 feet
(2) Medium: 200 feet
(3) High: 300 feet
Categories 1 through 4 assume use of package high service pump station.
Categories 5 through 12 assume use of custom-built finished water pumping station.

Abbreviations and Symbols

ac	alternating current
AGI	acute gastrointestinal illness
ASME	American Society of Mechanical Engineers
ASU	aereal standard units
AWWA	American Water Works Association
BAC	biological activated carbon
BLS	Bureau of Labor Statistics
°C	degree(s) Celsius
CCI	Construction Cost Index
CDC	Center for Disease Control
cf	cubic foot
CFU	coliform-forming units
C·T	concentration (mg/L) x time (minutes) to achieve a given percent inactivation (e.g., 99 percent) under defined pH and temperature conditions
cy	cubic yard
dc	direct current
DE	diatomaceous earth
ENR	Engineering News-Record
EPA	Environmental Protection Agency
°F	degree(s) Fahrenheit
FRP	fiberglass reinforced polyester
ft	foot
ft^2	square foot
ft^3	cubic foot
G	velocity gradient, feet per second per foot
GAC	granular activated carbon
gal	gallon
gpcd	gallons per capita per day
gpd	gallons per day
gpm	gallons per minute
gpm/ft^2	gallons per minute per square foot
GSFD	gallons of permeate produce per square foot of membrane

Abbreviations and Symbols (Continued)

hp	horsepower
HPC	heterotrophic plate count
hr	hour(s)
kWh	kilowatt-hours
L	liter
lb	pound
m^3/day	cubic meters per day
MCL	Maximum Contaminant Level
MG	million gallons
mg	milligrams
mgad	million gallons per acre per day
mgd	million gallons per day
mg/L	milligrams per liter
$mg \cdot min \cdot L^{-1}$	milligrams per liter per minute
min	minute(s)
mL	milliliter
mm	millimeter
MPNCU	most probable number of cytopathogenic units
μg/L	micrograms per liter
μm	micrometer
$\mu W \cdot s/cm^2$	microwatt-seconds per square centimeter
NIPDWR	National Interim Primary Drinking Water Regulations
nm	nanometer
NTU	nephelometric turbidity units
PFU	plaque-forming units
ppb	parts per billion
ppm	parts per million
psi	pounds per square inch
psig	pounds per square inch gage
Q_A	average flow
Q_D	design flow
RDA	Recommended Dietary Allowance
rpm	revolutions per minute

Abbreviations and Symbols (Continued)

SDWA	Safe Drinking Water Act
sec	second(s)
sf	square feet
SPC	standard plate count
T	ton
TDH	total dynamic head
THM	trihalomethane
T&O	taste and odor
TU	turbidity unit
USPHS	U.S. Public Health Service
UV	ultraviolet
yr	year

Metric Conversions

English Unit	Multiplier	Metric Unit
cu ft	0.028	m^3
cu yd	0.75	m^3
ft	0.3048	m
gal	3.785	L
gal	0.003785	m^3
gpd	0.003785	m^3/d
gpd/ft^2	40.74	Lpd/m^2
gpm	0.0631	L/s
lb	0.454	kg
mgd	3785	m^3/d
mgd	0.0438	m^3/sec
sq ft	0.0929	m^2